Edge of Survival

by

John Toenyes

and

Phil Scriver

Copyright 2010 Phil Scriver and John Toenyes

All rights reserved.
No part of this book may be reproduced or transmitted in any form or by any means, electronic or mechanical, including photocopying, recording, or by any information storage and retrieval system without written permission from the author, except for the inclusion of brief quotations.

International Standard Book Number: 1452885192
EAN-13: 9781452885193

J & P Publishing, Great Falls, MT

Cover design by Macjiggy Graphic Design

Photo credits
Photos by Phil Scriver: pg 9, 29, 50, 78, 134, 151, 182
Photos by John Toenyes: pg 38, 172, 188, 220, 256
Photos by Lynette Scriver-Colburn: pg 120, 256
Photo by Norman Anderson: pg 94
Photos courtesy USDA, Forest Service pg 18, 66, 224

Dedicated to all those
who participate in
historical re-enactments

Contents

Acknowledgements...7
Part I: Preparations for an Exploring Party
    Chapter 1: Setting the Stage..11
    Chapter 2: To Camp Dubois..19
Part II: Observations of a cook
    Chapter 3: Leaving St. Charles...Up the Missouri....................31
    Chapter 4: Entering Indian Territory..39
    Chapter 5: Winter at Mandan...51
    Chapter 6: Mandan to the Marias...59
    Chapter 7: Marias River to Canoe Camp....................................67
    Chapter 8: Canoe Camp to Beaverhead Rock............................79
    Chapter 9: Beaverhead Rock to the Nez Perce.........................83
    Chapter 10: Nez Perces to the Pacific Ocean............................95
    Chapter 11: Winter on the Pacific Coast..................................105
    Chapter 12: Fort Clatsop to Travelers Rest.............................121
    Chapter 13: Travelers Rest and Onward..................................135
Part III: Collecting and Evaluating the Information
    Chapter 14: Diet during the journey..153
    Chapter 15: Assembling the data..159
        --Some hints to consider
        --The military unit theory
        --Corn mills and coffee grinders
        --Dutch ovens
        --Personal equipment
        --Frying pans and cooking utensils
        --Those mysterious fire irons
    Chapter 16: Myths exposed...173
        --Indians taught Expedition
        --Nine pounds of meat
        --The Paul Revere legend
    Chapter 17: Field tests..183
        --Field test 6/28/04
        --Field test 10/3/04
        --Field test June-July 2005
    Chapter 18: Weighing the evidence..189
    Chapter 19: Formulating theories...199
        --Our theory of time
        --Our theory of additional equipment
        --Our theory of cooking methods

--Our theory on similar cooking
--Our theory on combined messes
Chapter 20: Completing the picture...215
--Short camps
--Long camps
--Winter camps
--Hunters and small groups
Chapter 21: In a nutshell..221
Appendices
Appendix 1: Recipes from the journals..225
Appendix 2: From the Corps Orderly Book..................................228
Appendix 3: Foods eaten by the Corps of Discovery...................233
Appendix 4: How it was cooked...236
Appendix 5: Pocket Soup, Portable Soup, Beefe Glue.................237
Chapter End Notes...239
Sources...244
Index...247
The Authors...256

Acknowledgements

In this book we have tried to give the readers the information needed for duplicating historic cooking of the Lewis and Clark Expedition. We are certain those who give it a try will have an interesting experience. To get the most out of such a cooking endeavor carefully consider where you "camp" while cooking up your food and where the food comes from. A camp in the back yard and food from the local grocery store will do, but the closer you can get to what the Expedition did, the richer and more varied your experience will be.

A hunting camp in the river breaks and cooking a deer you tracked for miles and finally shot adds a dimension that can only be present by your active participation. Do you roast it on a spit, boil it, or make a stew? In your back yard camp this decision may be very different than what you would do in your river breaks camp when you are tired from a long day on the trail. There you would do whatever is the quickest and easiest considering what you know and what you have to cook with. As Captain Lewis commented on several occasions, one of the basic and essential ingredients for a good meal on the trail was a good appetite.

This book is the result of an evolutionary process. We started to write about the Corps' food, equipment and cooking techniques. But as our research progressed we found ourselves delving into other elements to better understand the information we gathered and choices of methodology they had to make. We needed to get into the minds of the Corps to think like they did. The result was a book dealing with survival of a hunting and gathering group not much different from most of the other groups that lived in the lands the Corps journeyed through.

We left the text a bit unfinished in Part II where the "cook" examines the Corps' journals and makes notes on what he finds. The style is somewhat like the reader would find in a field journal before it may be taken and rewritten into a final product.

Because so little was actually recorded on the subject of food by the journal keepers our research included many related secondary subject, many of which we had to rely on others to give us direction.

The authors acknowledge the contributions given us by a number of people that greatly aided in our efforts to produce this book. Gene Hickman, Kenneth Wilk and Charles White guided us in understanding how the military was organized and operated at this period of history. The International Dutch Oven Society and Edward Quinlan, CeeDub Welch, Bruce Tracy, John Foster and Larry Jacobs made good suggestions on where to find background material on dutch ovens. Colleen Sloan spent time talking with us about dutch ovens and their history and possibilities for various alternatives. Her experience was invaluable.

Probably the ones that were the most impacted by our research for this book were the Lewis and Clark Honor Guard members who put up with our continual testing when we were supposed to be cooking their meals. They became the sometimes unwilling guinea pigs for our tests and theories. Hopefully their systems will readjust to the twenty-first century and they will appreciate the knowledge they gained.

We must also express our thanks and appreciation to our wives, Nancy and Barbara, who allowed us the freedom that we needed to pursue this project even though they may not have understood why and would rather have had us tending to the more pressing needs of keeping a house in repair and the hundreds of other ways normal families spend their time.

Part I
Preparations for an Exploring Party

This book is an indepth examination of how the Lewis and Clark expedition, the Corps of Discovery, fed itself. Instead of a twenty-first century cookbook with modern recipes for preparing foods eaten by the Lewis and Clark Expedition, we developed a book that is a detailed account of how the Lewis and Clark Expedition ate as they gathered at Camp Dubois then made their way across the North American continent and back. It is a study of what foods they took with them and what foods they found along their route and how these foods were prepared and eaten.

Chapter 1
Setting the Stage

You will journey into the unknown. To survive you must break the rules and make new ones as you go. Work as a team and extend the hand of peace to your traditional enemies. Carefully record what you see and do.

We commonly call it the Lewis and Clark Expedition, but it was so much more than that. As an expedition many people look only as far as the adventure story and compare it with other expeditions that resulted in immediate conquest or at least land acquisitions or blazing a new trail for settlers to follow. Detractors of Lewis and Clark rightfully say none of these resulted from their explorations.

A better way to consider the Lewis and Clark Expedition is to refer to it as the Corps of Discovery's journey. Think of it more as a well organized work detail that was given clear and distinct instructions of where to go and what to do. With this new mind set we can more easily get beyond the adventure story and examine who the Corps was and how they went about doing the job they were sent to do. One of the real values of the journey was how well the Corps did the job they were assigned.

The Corps of Discovery was a major departure from most anything that had been attempted before. They were well financed, well organized with a clear plan of action, and a military unit that was a well disciplined team. They were not sent to conquer a land, but to study it; therefore they would be in the field however long it was required to get their job done. Consequently they could not take all the required rations with them or establish a supply line for replenishment.

Jefferson had realized that he would need men that knew how to live off what the land provided; through their own hunting and gathering abilities and by their abilities to successfully trade with the natives they made contact with. The Corps of Discovery would by necessity live a subsistence life during their journey. That would mean they would have times of bounty and times of scarcity that may well severely test their ability to survive.

The Corps' journey becomes a study in survival and food played a major part in their success. Obviously people need food to eat or they starve to death. But food and survival goes well beyond that concept; subsistence living means knowing how to acquire food and how to make it edible. A person may have an abundance of food, but if he can't prepare it in such a way that it is edible he may well starve to death. Of course a person must be able to collect the food from the land. This means he needs to have the knowledge of what is edible and what is not and he must be a good enough marksman to kill the game animals he finds. Lewis recognized that fact when he asked Clark to recruit two good hunters. Clark took that further and recruited nine of the best hunters in Kentucky. He then spent the winter drilling the Corps of Discovery in marksmanship and care of their weapons so they were as prepared as was possible.

What foods are being eaten impacts how much work can be done. A person existing on a near starvation diet is not as alert or mentally sharp as one who has a more plentiful amount of food to eat. Food also impacts how well a person can tolerate the weather. For example a certain amount of calories are required for a person to withstand cold temperatures. The type and amount of food being eaten directly controls how much work can be done and at what rate. We see in the journals that the men's rate of work is adjusted depending upon how much food they are eating.

The particular route the Corps took was in part determined by their food. They were more able to tackle the rugged terrain of the river canyons in the prairies because of their better diet than while in the mountains when their diet was more restricted. Such a route restriction may create a lost opportunity for finding game that lay back from the river thus perpetuating a reduced diet. Sometimes a variation in their route was made to acquire food. Several times the Corps halted their march so they could resupply their stock of food. Hunters then had more opportunities to explore the area to find game. These side trips added to the Corps general knowledge of the area they were passing through. The location of Fort Clatsop was determined in a great part by the availability of a good food supply.

How the food is prepared helps determine how much is eaten. If food tastes better to a person, that person will tend to eat more than if the taste is poor or indifferent. Try building a fire out in the open and of sufficient size to cook a ten pound roast when it is raining and has been for the last three days. When presented with this kind of conditions an alternative food may be selected even if it does not have as high a calorie content or the quality of taste. Work rates may need to be adjusted to compensate for the difference.

Before we begin our work to discover how the Corps of Discovery fed itself we must put a few concepts into perspective. Our views today do not mirror those of the average person two centuries ago. Most people during that time were self sufficient rather than the specialists we are today. If they couldn't make it, raise it, or hunt/gather it they did without. Today we have others raise it, make it, or gather it then we buy it. One attitude we need to realize is what people thought of food then.

At the start of the nineteenth century most people were rural residents who raised most of their own food, supplementing that with what they could get by hunting. They only purchased a few items from the store in the not-so-near settlement. And that was only if what they needed was available, and if they could afford the price the store owner was charging. They didn't have the luxury of going to one of several close-by supermarkets if they ran out of some item, or if they decided a different kind of food would be good for a change. Today people make these trips almost daily and have a great variety in their diet, but in 1800 those trips were seldom and the variety came about only with the changing seasons or luck of the hunt. Consequently deer or bear hams for breakfast, dinner and supper for weeks at a time was commonplace. The meal may be completed with the addition of some corn, wild berries, bread and roots.

Depending upon exactly how far from civilization the individual lived his diet probably varied only by what was in season and by his abilities as a hunter. His was a subsistence diet which meant he lived off the bounties of the land, sometimes having enough to gorge himself on and at other times he starved. Consequently most people put very little emphasis on what they ate. The emphasis was on getting something to eat. Variety came only in those times of plenty. Today we almost demand a balanced and varied diet. In 1800 it was sufficient most of the time to get anything to fill the belly.

Only the very small percent of the population that lived in the few cities attached much more than the basic necessities to food and eating. Today eating is one of our common ways of socializing. In the cities during the past centuries this was also the case, but that was less than 10% of the population. The farther a person lived from the population centers the less important it became. This is one reason President Jefferson wanted people used to living on the frontier as members of the exploring party to the Pacific.

Another major difference in food preparation then and now is that there were no recipes for preparing food like we have today. The average person knew the basics of preparing the few foods he grew, gathered or shot. Each person developed a technique that served him based upon the situation he found himself in. These techniques were the quickest and easiest way to change raw food to an edible form. If a person could not hunt and/or gather the right edible plants well enough to secure his own food he perished. Likewise, if he could not prepare that food so he could eat it he perished. These people of the frontier did not spend hours over a stove preparing a meal; they seldom had the time or the desire for that. The whole activity of eating was not that important.

A second very important concept we need to understand is how people worked. The amount of work a person can do is dependent upon the amount and kind of food he eats. We see several examples in the Expedition Journals that testify to that fact.

A person that is accustomed to hard physical labor works at a set pace somewhat below his maximum ability, but hard enough and fast enough to accomplish the task. The rate of work is adjusted to allow the person to continue working for the duration of time he is required to be doing the job. A good example is a marathon runner does not attempt to sprint through the entire race like someone would do if he was running a hundred yards.

Instead he sets a slower pace that will allow him to complete the 26 mile course, but do it as quickly as he can. In 1800 most people were accustomed to hard physical labor whereas today most of us are not. When we are put into a situation of demanding physical labor we set a much faster pace than we can possibly sustain. Consequently we quickly tire and must take a break. The person who knows how to work within his range continues on without pausing to rest, and probably gets more done than the one who works faster but takes many breaks because he can't sustain the fast pace. We see this in the journals as the men worked to get the boats up the rivers. They adjusted their daily mileage depending on the total effort required to overcome the currents and other obstacles. As they got into the Rockies and closer to the point they had to quit the river their pace slowed even more because of the lack of sufficient food to sustain them.

Today we do not think much about whether we have enough food to sustain us during our physical labor, but to people who live on a subsistence diet it is a very real concern. Clark showed how very real this concern is during the winter at Fort Clatsop when they went to see the whale. He tried to help an Indian woman who was carrying a large load and had slipped while climbing a hill. To his dismay he could hardly lift her load which he estimated was only about a hundred pounds. He blamed his reduced strength on the Corps insufficient diet.

How does a small group of people prepare for a journey along some unknown route that will require an unknown length of time to reach some unknown place where they will stay for an unknown length of time then return by some unknown means? There would need be a certain amount of equipment purchased for transportation, clothing and shelter; but what foods and how much of it and what equipment will be needed to prepare it?

We started our quest to discover how the Expedition fed itself by carefully examining all that was recorded by the journal keepers; Lewis, Clark, Ordway, Floyd, Gass and Whitehouse. For this we used "The Journals of the Lewis and Clark Expedition" edited by Gary Moulton. Then related letters and documents as presented in the "Letters of the Lewis and Clark Expedition" edited by Donald Jackson were scrutinized. These comprise the body of primary materials written by the Expedition members at the time of their exploration. We continued, as is normal, with our research of the related documents. We studied the records of the Expedition from a standpoint of subsistence and the challenges along the way that created times of great bounties and times of near starvation.

There was very little written in the Journals about how the Expedition cooked their meals or the equipment they used for that cooking. This book was written to fill in those blanks with the best possible guesses we could make based upon our research and logic. Although we have included information about food and cooking during the Corps' winter at Camp Dubois our main interest is from the time the journey actually started in the spring of 1804 until their return to St. Louis in 1806. Foods and cooking techniques before the journey are probably very common of the time so very little was recorded in the abbreviated journal Clark maintained. In any event we know adequate supplies were readily available from the local farms and the merchants in St. Louis. Our interest focuses on how the Corps fed itself while on the move when their food had to be found and prepared in a shortened time.

With those documents completed our research turned to "hands on" testing. To test our theories we frequently duplicated as best we could what the Expedition did by building a fire and cooking some food over it. We tried different kinds and sized kettles, various cooking methods, etc. to come up with the most reasonable answers. In doing our testing we considered both short and long term activities. There may be multiple ways of accomplishing a task if it is to only be done once or very few times; most any way will do. However if that task is to be done (or could be done) frequently for an extended period, those methods and materials used tend to become more limited; consideration then must be given to time needed and available, the individual cook's abilities, what else is being done, etc.

After reviewing the story of the Expedition we collected the data we had mined from the written records and analyzed what we had. Using our test results and other bits of information that impact food preparation and requirements—such as weather, geography and the amount of physical work being done by the men—we drew our conclusions. The men of the Expedition did not eat much differently than they normally would have nor did they do much different tasks. The big change was that instead of doing what they normally did in civilian life as individuals, they did it during the expedition as a team.

Please bear in mind as you read this book that what you are reading is not entirely historical fact. Historical records are never complete. Over time some or all of the records may become lost or destroyed. Journal keepers never give a complete, accurate and unbiased account. If an event has multiple journal keepers there are almost certainly differences in what is recorded and conflicts are found. That is certainly the case with the journal keepers during the Lewis and Clark Expedition. To fill in the holes in the historical data several theories have been formulated by the authors as a result of our research and testing. We have tried to clearly mark those areas. Any opinions, ideas, conclusions or theories presented are those of the authors and we hope they are clearly marked as such. The intent was to use such terms as "we can conclude," "probably," "could have," "reasonable" or "logically" to set these apart from the documented facts. If any conclusions are not clearly marked as such we apologize. Please be assured it was an oversight on our part and not an attempt to pass off our ideas as fact. We do not believe that any author, regardless of stature, has that right unless the book is fiction or essays.

If additional primary documents are found sometime in the future, any or all of these theories could be proven right or wrong in any combination. Although we tried to be as thorough and diligent as we could be with the depth and breadth of our research, it is possible that we overlooked some item that could impact our conclusions. Only time will tell.

Some who read this will undoubtedly disagree with the theories we present—and that is okay. Our challenge to those people is for them to objectively research and offer a better conclusion.

Typical leather clothes the men of the Corps of Discovery replaced their uniforms with as they wore out.

Chapter 2
To Camp Dubois

While Thomas Jefferson crafted his plans for an exploring party to journey up the Missouri River and on to the Pacific Ocean, certain key ideas became firmly fixed in his mind. **(1)** These basic ideas would provide for the exploring party's survival. One of these concepts was that the explorers would be a military unit so the men would have the discipline, leadership and organization such an undertaking demanded. A small unit of one good officer and a dozen enlisted men would be sufficient. But this was not to be a normal military unit of its day. Unlike other military units that would spend a "season" in the field then return to garrison for the winter, this military unit was to be gone into the frontier wilderness until its mission was completed. **(2)** As we will see, the Corps of Discovery apparently was granted some flexibility concerning the written and unwritten military rules undoubtedly because of their unique mission.

Jefferson knew that the expedition he was seeking would experience many hardships before the exploring party completed their journey. Shortages of food would be one of them. He reasoned that since almost every member of the military had frequently experienced periods of short food supply (almost a normal aspect of life in the frontier forts), that experience and military discipline and training would become a key to the success for his party.

Another concept basic to the expedition was that the group's primary food supply would come from the land they were passing through. **(3)** Jefferson readily recognized that the exploring party could not carry a supply of food for the entire trip with them. He also knew and that once the exploring party departed from the Mandan villages there were no re-supply points along the route for them like the voyagers that crisscrossed Canada had. Consequently the Expedition must have the ability to live off the land. That meant hunting fresh meat from whatever sources that presented themselves as well as recognizing and eating plants--

roots, fruits, berries, leaves, etc. The other important food source was obtaining what foods they could through trade with the people they met during their journey. To help facilitate this food source Jefferson showed his genius when he carefully instructed Captain Lewis, "...in all your intercourse with the natives treat them in the most friendly and conciliatory manner which their own conduct will admit." **(4)**

Of course Jefferson was thinking of establishing a peaceful relationship with these natives for later commercial purposes, but he also knew some trading could take place at the time of the expedition's visit and obtaining needed food could be a part of that trade. The men Jefferson was sending into the vast unknown lands west of the Mississippi were not strangers to dealing with Indians. But this mission needed to be far different from what these men had previously experienced. Jefferson had to find a way to convince a military unit that had little, if any, diplomatic experience that these western Indians had to be met more as equals and encourage developing friendly relations if his dream of a commercial enterprise was to succeed. By tying friendly relations to successfully getting adequate food supplies, Jefferson ensured the Expedition had a very personal need to develop good rapport with the natives they encountered along their route.

The journals of the Expedition show that the Corps of Discovery was very successful in obtaining food from all of these planned sources except for those times when there simply was no food to be had. Lewis had prepared by purchasing quantities of Indian trade items in Philadelphia. Purchase orders show brass kettles, butcher knives, fish hooks and beads as some of the trade goods he bought. **(5)** He also specified the men who were selected to accompany him to the Pacific had to be used to frontier life and capable hunters (used to frontier life by necessity meant able to live off the land by knowing what plants to collect for food as well as being capable hunters).

Lewis' letter to Clark on August 3, 1803 shows more than just confirming Clark's status as co-leader of the expedition. **(6)** Lewis discussed the need to hire a good hunter or two and asked Clark to try to recruit some. Clark's response must have been total agreement and immediate action; he replied to Lewis on the 21st

telling him he had selected four of the best woodsmen and hunters in his part of the country. By the time Lewis met Clark at Clarksville/Louisville those four recruits had grown in number to nine; they are now referred to as the Nine Young Men from Kentucky. Several of these recruits proved to be more than adequate hunters.

The concept of a military unit in 1800, by design, living off the land was quite uncommon, if totally not against the prevailing military standards of the day. Units in garrison would supplement their military issue food with occasional packages from home and with locally purchased foods from time to time, but rarely through allowing the men to hunt or through foraging. **(7)**

Although there were often shortages of military issue food, most frontier camps and garrisons did not augment their rations by allowing hunting. Many of the officers were thoroughly opposed to having soldiers hunt under any circumstance. In 1809, for example, Colonel Bissell went so far as to issue a written order that forbad hunting unless the individual had a "special permit" (issued personally by Col. Bissell). **(8)** He further declared that the use of military guns and rifles for hunting was strictly forbidden. Those who might get a special permit had to supply their own rifle from civilian sources. We can conclude that the military ate the food provided by the Army and made due the best they could. Only occasionally would those rations be supplemented by packages from home or local food purchases.

The situation of unreliable food supply for the Army can be found during the Revolutionary War and continued well beyond the time of the Lewis and Clark Expedition. It was simply acknowledged that food supplies varied greatly depending upon the locale the Army was in. The Army's official orders provided for an adequate (although monotonous) diet for the men, but the lack of supplies frequently prevented "full rations" from being issued (These official orders were probably referred to by the soldiers in the field as dream sheets since they were seldom a reflection of what they actually had to eat). This was especially true during those times the Army was on the move and a reasonable supply system could not be devised that could keep up with the troops. Still the men were not permitted to hunt or fish to supplement the rations issued and all locally purchased rations had to be made through the Quartermaster and not directly by the men. In addition to regulating prices central purchasing provided for a more even distribution of the additional supplies to all the men and prevented the men from being taken advantage of by unscrupulous merchants. **(9)**

During the late 1700s and early 1800s military units did not have mess halls where the men were fed in large groups. **(10)** The men were divided into small groups of six to eight. Each of these messes lived together, slept together and ate together; cooking their food over an open fire in the mess area near their tent. When they were in garrison they cooked over the fireplace in their quarters. Rations were issued to the messes each day.

When the Corps of Discovery established their winter camp at Camp Dubois Captain Clark laid ground rules that broke most of these taboos on military men hunting to supplement their army rations. While Lewis spent most of his time in St. Louis gathering information and otherwise making preparations for the expedition's departure the following spring, Clark took charge of the camp and preparations there. The first order of business was to build huts for the men to live in. When completed, Camp Dubois took on the appearance of a typical frontier Army post.

Although initially conceived by Jefferson as only needing a dozen men, Clark was a bit more pragmatic in establishing the actual size of the exploring party. As the details of the journey were developed and the necessary supplies and equipment calculated he quickly realized a dozen men might not be sufficient

Clark spent considerable time during the winter at Camp Dubois calculating the amount of provisions and size of crew to go upriver as far as the Mandan Villages where the Expedition would spend the winter. From that point a smaller force would strike out for the Pacific while the balance of the crew would return to St. Louis. A smaller upriver crew would mean fewer provisions could be carried (The amount of provisions that would be needed was reduced even more by a greater dependence of living off the land). Clark had to figure the right balance between amount of food to carry and size of crew needed to get the job done. The two Captains determined they needed a larger crew than the dozen originally considered, but how many more men were required to insure getting to the Mandans since all the Indian groups between St. Louis and the Mandans were not considered friendly? The Captains finally settled on a force of about four dozen with 30 set to continue on to the Pacific the following spring. The group returning to St. Louis would be an adequate size to get the keelboat back with its cargo of notes, maps and specimen Lewis and Clark had gathered to that point.

As was normal at frontier forts for the commander to do when the men were in winter quarters, Clark contracted with local farmers and merchants for food and other supplies that were needed. He also spent much time transforming his men into a well-disciplined team that would be capable of long-term survival in the wilderness. A dozen years earlier he had learned the lessons of a junior officer well while serving under General Anthony Wayne. Very high on his list of requirements was for the men to learn how to take care of their weapons. A clean and properly adjusted rifle was essential for survival on the frontier. The men also had to be good marksmen, so target practice was frequent. A final basic skill that had to be so thoroughly a part of each man that it became second nature was skill as a hunter. Although most of the men knew the skills of hunting to some degree only a few were both adequately skilled hunters and marksmen. The men were taught tracking and the art of hunting. Not only was Clark developing a cohesive military unit from the men he had gathered from various other military units and recruited from civilian life, he was also assessing their abilities. While they all possessed the general qualities the President and both Captains wanted, most all of them had room for improvement.

Payoff for the hours spent drilling on care of weapons and target shooting was a trip to the woods where a good meal was the result of a successful hunt. During the winter at Camp Dubois hunting parties were out almost daily. Consequently most of the fresh meat was supplied by the hunters' rifles, with only a small amount purchased from the local farmers. Fresh meat varied from squirrel and rabbits to deer, bear and birds such as geese, turkey and swan, depending upon the hunters' success.

While the Expedition spent the winter of 1803-1804 at Camp Dubois they ate much like they would have if they were still at home (probably better than they would have in their old army units since spoilage of army rations causing shortages of food was common in the frontier forts). Since they were living in civilized country surrounded by farms and just across the Mississippi River from St. Louis, a bustling city of 1500 inhabitants, it would be fully expected that their diet would be similar to the normal population, or standard fare of a military unit in garrison with the exceptions of more fresh meat and no periods of food shortages. Once they left Camp Dubois on their journey up the Missouri their diet and eating habits started changing and constantly changed throughout their journey mirroring what the land produced.

Turning to the early pages of the Corps of Discovery's journals we do not find exactly what they ate or exactly how it was prepared. Words like 'provisions" or "rations" are found; but what those provisions were are not detailed. We can only assume a reference to standard military rations with possible augmentation of fresh meat and other local purchase items that are listed. We do get a partial list of foods, especially the fresh meat the hunters provided. Their meat consisted of a variety of animals from the woods and from the local farms. They also purchased potatoes, butter, eggs, milk, cheese, corn, bread, flour, vegetables, salt, sugar, turnips, lard and coffee.

We find no detailed descriptions of how the foods were prepared. Since they were eating foods well known to them and they were living in a "civilized" environment, their cooking methods probably were those typically used in frontier homes or military forts; stews and soups simmered in large brass kettles and roasts cooked on a spit over the open fire would have been common. We also are not told who is doing the cooking. We can only assume that it was a duty that rotated among the privates in each mess. Such a practice was common in military units. **(11)** Apparently in Clark's limited journal while at Camp Dubois no effort was made to record those things that were well known. This concept is frequently found in many of the journals kept by explorers of the west.

By April first the Captains made their initial evaluations of the men in camp, deciding who would be included in the group that was to go to the Pacific and those who would only go as far as the Mandan Villages then return to St. Louis with the keelboat. With these decisions made the Captains then built the formal organization of the Corps of Discovery.

In the organization we see the start of how the cooking duties were laid out. The Corps of Discovery was divided into groups of five to six men, called messes, with a cook for each mess which made the chore of cooking simpler. However, it apparently was a duty that rotated among the men. This organization and cooking duty was commonly found in the army units of that time. Consequently both Captains as well as many of the men who were already in the army were familiar with the organization the Captains laid out. (Please see Appendix II for a more detailed look at the organization.)

By the time spring arrived and the expedition set off up the Missouri River the routines of procuring foods from the land was well established. Although everyone was not an expert hunter yet, Captain Clark remarked they were all at least adequate marksmen and the group of men who were expert hunters was large enough to keep the expedition supplied with fresh meat.

As the Expedition traveled up the Missouri River their food was a combination of provisions they took with them supplemented by what the could get from the land they passed through and by trading with the people they encountered. To understand how well Jefferson's plan for feeding the Expedition by living off the land worked, we need to examine what provisions were taken from St. Louis and what was obtained along the way through trade and by hunting.

While the keelboat and pirogues were being packed for the trip upriver, Clark made lists of what was being taken. From these lists we see the foods were common military or frontier fare. Corn meal

- Coffee
- Hulled corn
- Sugar
- Flour
- Biscuits

- Salt
- Grease
- Beans
- Peas
- Pork
- Hog lard

Although these items were not on Clark's list we know from other journal references to them the Expedition also had a quantity of
- Butter
- Honey
- Portable Soup

Several purchases were made from farms near the river during the first weeks of travel including
- Fowl
- Various vegetables
- Eggs
- Milk

The volumes of provisions that were ordered were probably based on the Army's daily rations allowance. **(12)** In 1802, the ration for each man was 1 1/4 pounds of beef, or 3/4 pound of pork, 18 ounces of bread or flour, and 1 gill of whiskey, rum or brandy; each one hundred men should have received 2 quarts of salt and 4 quarts of vinegar. If we divide the ration by the number of men we find that the flour, corn meal or biscuits would only last between 5 to 6 months and the ration of salt about 19-20 months. It is obvious that the amount of provisions was far short of the needs of the Corps. Clark said they had 40 days provisions for the party packed on the boats ready for the trip upriver. He only wished they could be taking more.

As Lewis was readying for the trip to Philadelphia in early 1803 for his brief formal education and obtaining the supplies he would need for the expedition, the Purveyor of Public Supplies, Israel Whalen, was instructed by the War Department to get from private purchases whatever Lewis asked for that he could not get from public stores (military). While William Irvine, who oversaw the public stores in Philadelphia, was told to give Lewis what he asked for from the public stores. **(13)** Purchase receipts from merchants in Philadelphia show Lewis purchased these items that he apparently could not get from public stores. Since Lewis was not to be charged for items purchased from the public stores these bills to Lewis show they came directly from private businesses in the city.

- three dozen tin pint tumblers from Thomas Passmore
- two dozen tablespoons and eleven dozen knives (3 sizes) from Harvey and Worth
- three corn mills from Edward Shoemaker
- fourteen brass kettles, a black tin saucepan, and porterage from Benjamin Haberson & Son
- four dozen butcher knives from John and Charles Wister.

Because we have no lists of what Lewis got from public stores, we don't know if the equipment listed above was in addition to other items Lewis did get from that source, but since these receipts mirror Lewis' list of requirements, logic says he didn't get any cooking equipment from public stores in Philadelphia. This short list of equipment from merchants constitutes everything he got that is documented. His list was in keeping with what equipment would be needed for standard military cooking methods of that time.

When Lewis was making his purchases in Philadelphia he was equipping an expedition of a dozen soldiers, but the group that set out up the Missouri from Camp Dubois in the spring of 1804 was nearly four dozen. Where did the required additional equipment come from? Did Lewis make the needed purchases in St. Louis from private merchants there like he did in Philadelphia?

Part II
Observations of a Cook

Survival for the Corps of Discovery was not successfully defending themselves from Indian attacks, although in several instances that could have easily become the situation. It was not overcoming great obstacles presented by severe weather conditions or great natural disasters or wild animal attacks. The Corps' main survival encounters were securing adequate amounts of food that would allow them to perform the tasks required of them.

The Expedition only took 40 days rations so they had to primarily subsist on what the land provided. Consequently finding their food was a daily effort even in the prairies where abundant plant and animal life was available as they passed through.

Chapter 3
Leaving St Charles....Up the Missouri

It was mid afternoon on May 14, 1804 before the two pirogues and keelboat pushed off against the spring current of the Missouri River. Since it was far too late in the day to begin the expedition this was a ceremonial departure. Many last minute details were forgotten and men were running about. In fact, two of the men, Drouillard and Willard, were going to catch up in the next few days traveling overland.

I can imagine Clark was going over in his mind the last details of the procurement. Do I have enough rations for the men? Clark figured he had over 4,000 rations. Were they the right ones? He knew that the game on the land and the success of the hunters were the keys to their survival. He had trained the hunters to be sharpshooters during the winter just passed at Camp Dubois so the ammunition they carried would last the entire journey. He purchased cooking equipment but also knew that he had to rely on some personal equipment that each man carried with them. He knew the standard military rations were not going to make the entire journey-they only had a forty-day supply and they would have to trade with the Indians and perhaps purchase food items from trappers they might encounter on the way. As it began to rain, the Corps made it to the tip of an island only three miles above their departure point. Clark in all the excitement did not dispatch any hunters during the day and relied on rations for the evening meal. As they continued the next day, they came upon a small camp of Kickapoo Indians. These Indians had evidently made a prior trade agreement to supply the Expedition with deer meat in exchange for spirits. The fresh meat substituted for the daily meat ration for each man.

The keelboats were cumbersome and progress was a laborious chore, although they made use of a sail when the wind direction was right. They passed several small farms of American settlements and on May 23 Clark sent the Field brothers to purchase some corn and butter. Clark knew that each day he could supplement the rations from the land they passed through the longer they would last. Willard and Drouillard finally caught up with the crew on the 4th day. They had shot a deer, which was only about 30 to 45 pounds of edible meat; the rest of the animal was bones, hair, horns, intestines and other parts not fit to eat.

Lewis discovered they were consuming the provisions faster than he had planned. He needed to organize his messes and have a system of rationing. They camped at a French village and purchased some milk and eggs. On May 26 Lewis presented a Detachment Order [see appendix]. He divided the men into three messes under the direction of Sergeants Floyd, Pryor and Ordway. The order also broke the French detachment into two messes under DeChamps and Warvington. The men that formed these two messes consisted of the crews of the red and white pirogues. The sergeants were exempt from making fires, cooking and pitching tents. The rest of the mess was to perform an equal portion of the duties. The provisions were issued each evening before cooking. A portion of the provisions was reserved for the next day. Cooking was not allowed during the day while on the march. Ordway had the duty of issuing the provisions to the messes. The first day's rations was dried corn and grease, the next day Pork and flour, the next Indian corn meal and pork. When fresh meat was available, pork was not issued.

The rations issued followed the military recommendations of the day. Each man was to receive 1 ¼ lb of beef, ¾ lb of pork, 18 oz of bread or flour and 4 oz (1 gill) of whiskey, rum, or brandy. Each 100 men received 2 qts of salt, 4 qts of vinegar, 4 lbs of soap and 1 ½ lbs of candles. **(1)**

Lewis had procured several items that were not common military rations of the time. [See chapter 2]. The ones that stand out are the 600 pounds of grease and 12 kegs of hog lard. He also had 100 pounds of beans and peas. Beef and vinegar were not on the list of provisions taken.

As they proceeded on, Lewis began to send more hunters out to supply them with fresh meat. Their hunting success was poor, as the next few days only secured two deer. On June 1, they made camp on a point of land where the Great Osage River converged with the Missouri. Lewis planned to spend a couple of days there to make observations. Whenever the march stopped the work of preparing and storing foods for the next days march began. On June 2, Drouillard and Shields returned exhausted, after seven days hunting. They were able to secure four deer. Because few deer were brought to camp, there probably was very little remaining for drying. The next few days the hunters were more successful. They killed seven deer on June 4 and the men "jerked" or dried the meat. Drying the meat, or removing most of the blood, allowed it to keep a few more days. They set out early in the morning and had little time for the sun drying process. Sometimes the process was hurried and instead of drying it in the sun, the meat dried above the heat of the fire to hasten the process. Having dried meat the men could carry with them gave them food for the next day.

Since they did not take vegetables, except for the dried beans and peas, the men were constantly watching for edible plants. On June 5, Lewis spotted some watercress along the shore of and island and had York swim to the area and collect enough for supper. During the next few days, buffalo signs were present and Drouillard killed three bear, a sow and two cubs. These were black bears, which were a common meat on the frontier. Clark does not say anything about eating the bear, but it would be my guess that they enjoyed the change of menu.

Cooking methods were not discussed in the journals. Cooking seemed to be a simple process that did not take a lot of time. Roasting meat takes time so when they were on the move they probably did not roast much meat. However, because the military required boiling, probably most of the meat was boiled. Perhaps Lewis' procurement of 600 pounds of grease was used to boil the meat. It would make sense that because boiling water only achieves a temperature of 212 degrees and boiling grease would get to 350-400 degrees, boiling the meat in oil would take less than half the time needed for boiling in water. Grease and lard seemed to be very important commodities. On the June 12 Lewis purchased 300 pounds of buffalo grease and tallow from a voyager heading down river. The quantities purchased suggest that the grease was used for more than mosquito repellant. Only a few days into the journey and already they have purchased over 1,000 pounds of grease and lard.

The river current ran strong with numerous dangerous sand bars. The hunters were rather successful the last few days in killing deer and bear. The work of the river boatmen was very strenuous requiring the men to eat several times a day. The current rations given to the men were not adequate. The party was getting boils and dysentery. The captains blamed the water, which was dirty and contained a scum. Few vegetables were available so the diet consisted mainly of meat and dried protein (beans and peas). The meats they ate contained very little fat except for some of the bear meat. Boiling their meats in grease may have given them some quantity of fat but probably not the amount necessary to sustain their strength. Clark writes that two thirds of the men had boils. The mosquitoes were constantly disrupting the men's work. They were using the grease as a repellant. I cannot imagine how grease kept the insects at bay but they thought it worked. However, the men had boils, which is a staphylococcal infection of the hair follicle. **(2)** The grease may have made the boils worse.

Bacteria, unknown at that time, would be a major problem as they proceeded to the Pacific. The water that churned up from the spring run off was muddy and contained many bacteria and single cell animals that caused severe cases of dysentery, fever, cramps, and muscle aches. Clean water probably was not available at these lower elevations.

On June 23, Clark got off the boat and walked on shore. He planned to meet the boat on the next bend of the river but a strong wind prevented the boat from advancing that distance. Clark shot a deer and camped for the night. At about dark Drouillard caught up to him with the horses. They remained in camp that night and ate roasted deer meat. Without cooking equipment, they probably used green wood sticks to roast their meat. Clark and Drouillard returned to the boat in the morning with a bear and two deer. Several other hunters were having great success in harvesting deer and bear. In the evening, they jerked the extra meat.

On June 27, game was very abundant so the Captains decided to spend a few days at the mouth of the Kansas River. Eight to ten hunters went to collect as much meat as possible. The men unloaded the boats and dried their damp and wet articles. They found several personal items were ruined. The men were employed dressing skins and were able to take time to relax. Clark commented that this area would be a wonderful area for a fort. John Collins and Hugh Hall tapped one of the whiskey barrels and were court martialed. They received 50 lashes each. It is evident that the rations were under constant watch and the party took it seriously when the rules were broken. Lewis had taken 120 gallons of distilled spirits. At the rate of four ounces per man a day, that would last about 104 days. **(3)**

Clark commented on June 30, that they opened a bag of bread; he was probably referring to biscuits, which keep for a long time. They also opened some bacon and found it in good condition except for the bacon purchased in Illinois. It looked spoiled but Clark said it still tasted good. They were now in the beautiful prairies where they found abundant deer, raspberries, grapes and on occasion fresh spring water. Fresh fruit was probably a great hit with the men. I can imagine the sweetness of the raspberries were a great treat.

July 4 started with the sound of the swivel canon. Even though it was Independence Day, they proceeded on, working their way up the river. As they settled in to camp that evening, each man received an extra gill of whiskey and the cannon was fired.

Lewis determined that the distribution and use of the rations could still be better. He seemed to feel that "too many cooks spoil the broth" theory probably was not the best-organized ration control. He put one man in each mess responsible for the consumption of the provisions. The menu for each meal was the responsibility of the appointed man, and he was to use his judgment in its preparation. These three men Privates Thompson, Collins and Werner, had the job title of "superintendents of provisions." Though they had previously been court-martialed and disciplined, Lewis seems to have confidence in them. Perhaps Lewis and Clark felt this was a good way to keep an eye on them, or else their confidence in these men has now grown. Giving each man the flexibility to be in charge of their mess and allow them to exercise their own judgment on the mode of cooking, in my opinion, was the first step down from the military methods which kept standardized procedures. Lewis could see that those standardized procedures were difficult and undoubtedly felt that this was a way to provide each mess with a little flexibility. The men of each mess probably now felt that items such as fruits and vegetables and special catches such as a beaver, could be utilized in their mess rather than sharing with the others. The mode of cooking also allowed the Superintendents to use a mode that satisfied the men and the mission but not necessarily the strict military methods. Boiling meat in water is a slow and tasteless method. This may have allowed for more roasting and frying. On July 12, 1804, the party much fatigued, rested. Ordway noted that they rested and washed their clothes. Clark remarks, "I got grapes on the banks nearly ripe, observed great quantities, of Grapes, plums Crab aples and wild Cherry, Growing like a commn. Wild Cherry only larger and grows on a Small bush on the side of a Clift Sand Stone." The hunters collected and ate fresh fruits as they traveled. They may have collected quantities for their mess for the evening meal. Willard was court-martialed for sleeping on his post and was sentenced to 100 lashes.

On July 17, Goodrich, known as the best fisherman on the expedition, caught a large catfish. The hunters had continued success so each mess had enough meat to please the men. Goodrich probably utilized the Catfish for his mess. Catfish is an easily cooked fish either in boiling water or oil. Filleted fish, dipped in corn meal and boiled in oil makes a delicious meal. This was probably a delicious treat for the men, perhaps with fresh berries for dessert.

July 19 Clark writes that he had breakfast of roasted deer ribs, with a little coffee. It intrigues me to see the Captains now roasting meat. Clark seems to enjoy his meals as he now begins to describe some of them. Clark went ashore and found an endless prairie. Drouillard still was the most successful hunter and seemed to bring meat to camp daily. Clark talks of a yellow wolf he killed. I doubt that the meat was consumed.

On July 22, at the mouth of the Platte River, Lewis wanted to stop a few days and take some readings. The hunters were still having great success, so the men rested and jerked meat for the trip ahead. Some of the damp articles were lain out to dry in the sun. On July 29, Clark noted that the men caught three very fat catfish. The catfish were so fat that a quart of oil came out of the fish. I would imagine that while boiling the catfish in water the oil from the fish would float to the top. Clark commented about the large number of catfish. I would imagine the men dined in each mess on the freshly caught catfish. As kettles of oil boiled over the evening campfires, the freshly caught fish were cooked immediately. Turkeys and geese that had been shot were cooked along with the catfish. Clark commented that the men were in high spirits. Perhaps now they were more satisfied with a more varied and adequate diet.

August 1, 1804 was Clark's birthday and to celebrate he ordered a saddle of fat venison, an elk fleece and a beaver tail to be cooked and a dessert of cherries, plumbs, raspberries, currants and grapes. To prepare a beaver tail, which was a favorite, they would blister the skin using a rack of willow over hot coals then remove the skin after it was cool enough to touch. The tail was then boiled in water or oil. The meat looks like a fatty cartilage material and when cooked has a sweet flavor not like the rest of the beaver. The saddle of venison is the meat lying between the top of the hind legs and the first ribs. This is the filet mignon of the deer. It was probably roasted and brushed with grease for flavor, as venison is a dry meat. An elk fleece is a steak cut from and elk. The desserts were the ripe wild berries, with perhaps a little sugar or honey added to take away the tartness.

Drinking the cool, clean, pure water from Giant Springs was a marked contrast from the dirty waters of the lower Missouri. Clark blamed the water for men's boils and dysentery during the early part of their journey in 1804.

Chapter 4
Entering Indian Territory

The Corps of Discovery now entered Indian Territory. Game animals were in great abundance so the hunters were having great success. As the Expedition got closer to more heavily populated areas, the game once again became scarcer. Clark commented that four quarters of a very large deer weighed 147 pounds. It was so fat that it had an inch of fat on the ribs. Drouillard and Colter brought elk to camp on August 2.

Clark continued to write in his journals, but Lewis seemed to write with science in mind only. He did not talk about the daily activities, but described the different species of animal and plants that he observed as they pushed up the river.

That day they encountered a party of Otos & Missouris Indians. They met with six chiefs and exchanged roasted pork, flour and meal for watermelons.

As they moved along the favorite foods of Clark begin to appear. If he saw turkeys, he would usually kill one for dinner. He also always mentioned beaver tail. They cut the breasts of the turkeys boiled them in oil. It would take only about 10 to 20 minutes to cook a turkey breast in oil. Roasting a turkey would take much more time and it seemed that when they returned to camp they wanted to eat as soon as possible.

Clark does not seem so concerned anymore about the dysentery and fevers of the men. They were finding better sources of purer water than the Missouri was although the spring runoff was over so the river ran much cleaner. He frequently speaks of springs or clear streams.

On August 12, the Captains selected Pvt P Wiser to be the Superintendent of Provisions of Sgt Floyd's mess. There was not a reason given for Pvt Thompson's removal from that position. One could speculate that perhaps the job was not up to standards, or he was not making good use of the provisions.

Game became harder to find the next few days. The Corps passed the ruins of an Omahas Indian village. The tribe apparently contacted small pox that wiped out most of the members. They were a horticultural tribe that raised corn. To Clark's amazement they had gone inland to hunt. This could explain the shortage of game in the area.

On August 15, while searching for food, Clark took ten men and built a brush drag. They pulled it across a beaver dam and caught 318 fish. The fish were pike, brook trout, gold eyes, bass, red horse, catfish and a freshwater drum. While the men were using the brush drag Clark collected some fresh water shrimp and clams. He commented that the flavor of the shrimp was much like those of the shrimp at New Orleans and the lower Mississippi. Of the large number of fish caught, they probably ate only the best. It would not take long to cook these fish using boiling oil or for that matter water. Fish dipped in corn meal, and boiled in oil, regardless of the type, are a tasty change especially in the absence of larger game.

While they were camped at the Fish Camp, in northern Nebraska, for a few days rest Lewis went back to the same area with 12 men and collected about 800 fine fish. In the catch were 79 pike, 8 trout, 1 rock, 1 fat back, and 127 buffalo and red horse, 4 bass and 490 catfish and many drums. They captured Reed, who had previously deserted and sentenced him to 500 lashes, and then removed him from the permanent party.

August 18th was Lewis's birthday but unlike Clark, he did not procure any special meal for the day. However, the men received an extra gill of whiskey and danced.

On August 19, Sgt Floyd became violently ill. Clark was very concerned and had York tend to him. Floyd doubled over with extreme pain in his lower abdomen and could not keep anything in his stomach. They were going to give him a warm bath hoping to relieve the pain, but he passed away before they could get him to the bath. Before he died, he asked Clark to write a letter for him. They buried Floyd with full military honors on a ridge overlooking the river, and put a marker with his name and date on his grave. It is believed Sgt Floyd died of a ruptured appendix and peritonitis. **(1)** Floyd's grave was near present day Sergeant Bluff, Iowa.

On August 22, the Captains took a vote of the men to select who would replace Sgt Floyd. The decision was between Bratton,

Gibson and Gass. Gass had the highest number of votes. Two hunters left on horseback to seek game and did not return that day. On the 23rd, Clark commented on shooting a fine buck and Joseph Field killed their first buffalo. Lewis and 12 men moved it to the river to butcher it. Later Ruben Field arrived with two deer and Collins killed a fine doe. Lewis shot a goose and one of the men caught a beaver. They had fresh meat again. The wind was blowing so hard there were clouds of dust. They camped early and cooked some of the meat.

With the three messes, each cook prepared their meals. Due to the winds howling, the fires were far too hot for roasting, but getting their kettles contents to a high temperature with either oil or water would have been easy. With the addition of the daily ration of corn or flour, the evening meal was satisfying.

The next day Clark commented that a berry resembling a current, probably a gooseberry, was of a fine quality and would make a delightful tart. It is quite striking that on a journey of this magnitude that Clark would be thinking of making a tart, although a tart is very easy to make and would have been an excellent dessert. All that is necessary is a pot with about 1 inch of sugared berries covered with a patty of flour, water and lard; about a one to four ratio of lard to flour. Cooking with the kettle surrounded by coals, to keep at a medium heat, a heavier pot with a good fitting lid such as Dutch oven would be necessary. Tarts were a common dessert.

On August 24, Lewis, Clark, Drouillard, Ordway, Shields, J. Field, Colter, Bratten, Cann, LaBiche, Warvington, Fraser and York went to visit the Mountain of Evil Spirits. The group was gone most of the day. During a rest stop along the river, the men collected large numbers of different berries and probably took them back to camp. They ate jerked meat and two "Prairie Larks" or probably meadowlarks. **(2)** It seems to me those birds would have been only a mouthful each.

The men continued to jerk meat any time they had a few hours to spare. This meat is obviously not as dried as we think of jerky but probably they removed the blood, as time did not allow for thorough drying. When they said, they cooked jerked meat the cooking probably was done in either oil or in water. Dipping the dried meat in oil then rolling in corn meal, and dropped in boiling grease, would be delicious, and similar to a chicken fried steak.

Lewis made it official on August 26, and wrote an order officially appointing Gass as the Sgt in charge of Floyd's mess. Hunting was good the last few days and a lot of meat was jerked. Pvt Shannon has been absent for several days and the hunters, Shields and Joseph Fields were sent on horseback to look for him.

On August 27, Lewis set a prairie on fire to attract the attention of the Indians and soon some young Indian boys appeared and told the Captains that a Sioux encampment was up the river. Lewis prepared a speech for the Indians. He also wrote an order requiring the crews of the pirogues select a man to cook for each of the two messes that they had. I am not sure why the waited for a selection of cooks other than the fact that the Frenchmen again were complaining of the poor rations that they were receiving.

On August 29, about seventy Sioux arrived and camped across the river from the Corps. Lewis sent a couple of men to their camp with a bucket of corn and two twists of tobacco along with six kettles for the Indians to cook the game in that they had killed. When the men arrived, to their amazement the Sioux approached with a buffalo robe to carry Lewis or Clark to their camp. Neither being in the party, the men proceeded to the Indian camp and found a feast of dog roasting on the fire. The Sioux presented the fat dog as a gesture of their respect for the party. Of course, they ate and commented on its good flavor. Across the river, Ordway commented on the abundance of catfish, their size, and good taste.

At 10 O'clock, the next morning the Sioux arrived at Lewis' camp. Lewis gave his prepared speech and in the evening, the Indians danced around the fires. The Chiefs all gave their speeches during the next day and received presents. On 31 August, the Sioux returned to their camp.

As the Corps moved up the river, the journal keepers described the cliffs along the river and they observed the ruins of several Indian villages. There was not much written about meat in the journals except on September 2, Clark noted that Newman and Howard had killed four fine elk and all of the meat was jerked. They were catching catfish in this area and undoubtedly ate the fish instead of the meat. Colter was in pursuit of Shannon who was ahead of the boat.

Many new animals began to appear as the Corps made its way up river. Mule deer, prairie dogs and antelope were among the elk and buffalo herds. On the eighth of September two buffalo were killed along with a bull elk, a fawn elk and three fawn deer, three turkeys and a fox squirrel. Most of the meat collected was dried, but I speculate that the Captains had turkey for dinner. The squirrel was for identification purposes. These plains at this time of year were full of animals. The hunters were not having any trouble supplying the Expedition with enough meat to feed them well. On September 9, Clark mentioned that York killed a buffalo at his command, but that he could not get a black tailed deer.

On September 11, Shannon finally appears, half-starved. He had been out for 16 days and he stated that he had only a rabbit and berries to eat. He had run out of bullets. Clark commented that without bullets and powder a man would starve in a land of plenty. Shannon being the youngest of the troop was gaining experience the hard way.

Lewis commented on September 13 that he had killed a porcupine that was feeding on a cottonwood tree. He said, "The flesh of this animal was a pleasant and wholesome food. Skinning this animal is a laborious task but once the flesh is clear, it is a good tasting meat. Lewis was collecting many animals and preserving them so that he could send the specimens to Jefferson. On September 16, Lewis describes his first Pronghorn.

Fall was now upon them and the temperatures were beginning to drop. Clark issued a flannel shirt to each man and powder to those who had run out. Lewis and six hunters went ashore and walked over the countryside. He saw several thousand buffalo and many other animals spread across the great plain. They were in an area of abundant food. He describes plums that were ripe to eat and were of the same wild quality that were in the east. They were probably even the same species. **(3)** Lewis describes his dinner that was packed in his possible bag for the day, dried meat and a half of biscuit. Plums were probably eaten as they were ripe and ready. He drank clear, cool water from a pool left by the rains the day before.

Clark and some of the crew dried their cargo that had gotten wet from the rain and damaged some of the items. Of course, when they rested they collected and jerked as much meat as they could. They now are finding that the catfish are getting harder to come by, but the larger animals; buffalo, antelope, deer and elk are present in large numbers. On September 23, some young boys of a Teton Sioux village ahead warned the Corps of eighty lodges ahead on the next bend. Early on the 24th the Corps of Discovery prepared for the visit. They prepared items of clothing and a few metals for the Chiefs.

Things did not start out real well with the Sioux. Colter watched one of the Indians steal a horse and reported it to the Captains who now have an even tenser situation than just their arrival. He met with one of the Chiefs at the Teton River, but he was not aware of the theft. The Grand Chiefs called a meeting for the following day. It appeared that the hunters stayed within sight of the boats but still managed to obtain game for the camp that day. Ruben Fields shot a doe antelope and Colter shot several elk. Large herds of buffalo were seen farther inland raising large clouds of dust.

The day came for the meeting with the temperamental Teton Sioux. Their reputation of being thieves, beggars and demanding large gifts for favors now is forefront on the minds of the Captains. Jean Baptiste Truteau warned the Captains of these Indians and told them they should avoid them. The Captains learned before they left that this was an important threshold for trade to the Pacific as the Brule's tribes were very powerful and controlled trade up and down the Missouri as well as trade to other tribes in the area. Consequently, this meeting was of the foremost importance. **(4)**

Guarding and watching replaced hunting and fishing. They had camped on the opposite side of the river for protection and prepared for the meeting. Of course, as most meetings begin, one would try to calm the tension with food or drink. As the Indians offered several hundred pounds of fat buffalo meat and Clark in return offered cured pork to eat. They smoked a little then he served a little grog, which later became a problem.

Not having a good interpreter to translate all the words

properly things began to get difficult, and as the alcohol began to take effect things got out of hand. The Indians were not satisfied with the presents and demanded more. Clark finally reached a point where he had to draw his sword. Lewis then ordered everyone to take up arms and they retreated to the shore where the boatmen prepared to leave, but the Indians took up the rope to prevent their departure. The Indians had their bows drawn. Clark spoke with positive terms and two of the Indians waded along side the boat and Clark took them in and proceeded up the river to the camp.

Early the next morning, the Indians lined the bank of the river to see the boat. Lewis, at the request of the two chiefs, went to shore taking five men with him. In the evening, the Indians carried Lewis to a feast on a buffalo skin. In the Indian council tents, they honored Lewis with the roasting of a ceremonial dog along with about 400 pounds of buffalo meat over a large fire. The Indians also served dried meat pounded to a fine ground texture to which they added marrow from buffalo bones then made into patties. Clark did not enjoy the "raw" dog that the Sioux thought was a great delicacy. He also described the boiled white apple cut into chunks the size of hominy.

Ordway described a delicious soup made from the white apple. **(5)**(Psoralea esculenta Pursh). The white apple is a bulb found on the prairies. The collected white apples were threaded onto a cord then dried for later use. The Indians pounded the roots into flour and when they added water, buffalo fat, and bone marrow would make a tasty soup. In the fall, boiling dried meat and fresh bulbs make a tasty and healthy stew. Although the bulb has little taste itself, with a little salt and fat it comes to life.

Several Indians took some of the meat and pemmican to the Expedition's camp for the men then returned to their own camp bringing Captain Lewis with them. He arrived on a robe just as Clark had. The Indians then killed several more dogs and prepared them as a mark of friendship. Dancing carried on until a late hour then the captains returned to camp. The next day several chiefs dined on board with the captains as some of the men went to the Indian Camp.

On September 28, the Corps proceeded on. The Captains

fatigued from their experiences, moved up the river far enough to get away from the Indian village and camped on an island in the middle of the river. They cooked, probably some of the fresh buffalo that the Indians presented to them and rested well. The next morning they pushed on up the river still followed by a few of the Indians. Tension still lingered as Indians were watching all along the shore. On 30th of September, the Teton Sioux finally were behind them.

They continued up the Missouri struggling over many sandbars. The river was running low during the fall months and had become very hard to navigate. October 3rd, Clark describes damage to the goods from mice that got into their reserves of corn and scattered it about contaminating it with dropping. Some of the extra clothing had holes chewed in them. Indians still were being seen everyday. Refusing to talk, they proceeded on.

There were no journal entries written about hunting during the next few days. I would assume their fresh meat had nearly dwindled and they may have been eating dried meat or some of the preserved pork. The absence of game was the result of a large population of Indians living in the area.

During the next few days, game again started to appear. The hunters killed four antelope and a buck deer. That evening after dining the captains refreshed the men with a little whiskey. The Corps passed abandoned Aricara villages every day. When the game becomes scarce, the Indians moved camp.

On October 6, they stopped to examine an abandoned village and found three different kinds of squash; probably Duchesne, Hubbard and Turban. **(6)** Ordway wrote that they collected many squash and a few baskets left behind by the Indians. This would be a great diet change, fresh yellow squash. They killed an elk for dinner. To cook the squash would have been an easy task. Since burying and roasting took to much time to prepare, they probably boiled them in water until they were soft and palatable.

They continued up the river with moderate success at hunting and on October 8, approached a village of Aricara Indians. The Aricara appeared to be good farmers. On the island where they lived, they raised squash, tobacco, corn and beans. These were friendly Indians. After visiting the Indian village, the men learned that they supplied many other Indian bands with food and traded with French traders. They gifted the Corps with corn, beans, dried pumpkins, squash and other cultivated products. Clark commented that these Indians were not fond of alcohol.

A meeting with three Chiefs of the area went very well and this peaceful tribe welcomed the Corps into their community. Over the next couple of days, they visited three villages in the area where they were welcomed and gifted again with bushels of corn and beans. They received bread made from corn along with large beans called hog peanuts or ground bean plant. Meadow mice collected the beans and stored them in their burrows. The Indians took the beans but replaced the beans with other types of food so that the mice would still have something to eat. **(7)**

On October 12, Clark visited with the chiefs of each village. The Indians wished to make peace with the Mandan tribe. They asked the Captains if they could send one man to the Mandan village with the Corps to make the peace. Lewis agreed and the Chiefs gave the Corps 17 bushels of corn along with beans, tobacco and squash.

The Aricara lodges were made of earth. Wooden poles 30-40 feet in diameter with the roof supported them. Willows and grass laid across the roof supported a sod grass outer covering and a five to six foot hole in the middle for smoke to escape. These shelters seem more stable and warmer than the hide teepees that the Sioux used. They were a proud people that are poor, kind, and generous. They do not beg and give with great pleasure. **(8)**

October 13th John Newman was court-martialed for verbally attacking the military Corps and creating hate and discontent among the troops. They found him guilty and sentenced him receive seventy-five lashes and disbarred him from the permanent party. He was assigned to the crew of the pirogue.

The weather turned wet the next few days. On the 15th thirty Indians joined them bringing meat to exchange for fishhooks and beads. The Indians were curious and very friendly. They provided several meals for the men, visited, and smoked with them. York was a big hit with the Indian women and they were very fond of the other men.

The reduction of hunting pressure from the Indians makes the animals more plentiful. Clark watched the Indians kill antelope with sticks as they crossed the river. Large herds of buffalo, elk and antelope wandered with in eyesight of the shores of the Missouri. The Chief informed Clark that the animals return in the fall to the Black Mountains where they spend the winter.

Fall was fast approaching; the temperatures were dropping in the evening and the leaves on the trees were rapidly falling. The water level in the river was dropping and many sandbars appeared. The water was very salty and not very palatable. The men killed many animals as they moved rapidly up stream. Cooking the meat had to be done quickly in the evening. Boiling the meats in oil and drying as much as they could around their campfires was still the evening chore.

Not much fanfare or special cooking probably took place. They did not talk about any special meals or of breads or vegetables although they may have boiled some squash that may have still been edible. The hard-shelled squash kept very well but the yellow thin-skinned squash spoiled rapidly. They would have been using some of their recent gifts of corn but conserving most of it for the upcoming winter.

The country is beautiful and animals are plentiful. On October 20, Cruzatte wounded a white bear and was so startled by the animal's size and ferocity he lost his gun in his rapid exodus. This was their first encounter with the Grizzly bear. The Mandan villages were only a few days ahead.

Winter was approaching. On October 12, Clark commented that he had killed a fat buffalo with his small bore rifle, probably a Kentucky small caliber rifle. It snowed on the 21st with the temperature dropping fast. As they proceeded up this almost woodless area of the river, Indians watched them throughout the day. The Corps fought the endless sandbars of the lower Missouri. Timber began to appear on the shoreline of the river. They passed island bottoms where the Indians cultivated vegetables.

On the 26th they arrived at the first Mandan town. Villagers interested in the Corps arrived to welcome the Captains and crew. The Captains impressed the Chiefs with a steel corn mill. As the party moved northward the next few days they met with the Menetares, Gros Ventres, and Hidatsa. They received gifts, smoked and made peace.

Winter was now upon them and they needed cover for the winter. Clark looked upriver for seven miles for a place to build a fort, but he found that the river to the north had insufficient timber so he decided to return south to the next point where there was timber to build.

Sgt Charles Floyd was the only member of the Corps of Discovery to die during the journey. His grave is marked with the obelisk (below) and plaque (left).

Living off the land also meant surviving Mother Nature's harsh weather. The Captains had heard from traders in St. Louis that the winters at Mandan were sometimes very cold and they knew that they were running out of time before winter set in. They did not waste any time. On November 3, 1804, they commenced building their shelter for the winter. By befriending the Indians of the area and trading with them, they procured rations for a few days at a time. Hunters immediately went out with a pirogue to find fresh meat. It was soon obvious that the large populations of Indians in the surrounding villages were also hunting which meant fresh meat would be hard to find. They would have to range farther away to get an adequate supply of meat. Their supplies of corn would be a supplement but would not be enough to carry the populations through the winter.

Sergeant Gass probably had a major part in construction of the Fort. He described it as a triangular stockade, with two converging rows of huts. A gate at one end and walls were 18 feet high. Fireplaces made of stone were the heat source and probably used for cooking. Somewhere in the fort was an area to store fresh meat as well as squash and corn. One advantage of winter was that the fresh meat froze on the cold days so it did not spoil as quickly as it would in the summer months. The result was much longer shelf life with very little waste. The Indians were masters at drying their crops and storing them for the winter. Dried squash and pumpkins were sweet and were a good trade item to the Expedition. The captains knew the value of the traded food and traded as often as they could. They may have stored the products for use when they headed upriver in the spring.

The Indians also worked very hard collecting meat for the winter. Parties pass by the Fort everyday to hunt. On November 5, they drove a small heard of antelope into a brush pen and within two days caught 100. The intensified hunting was driving the animals further and further from the Fort. Some hunters who had gone downriver in the pirogue had been gone for more than a week without word from them.

The shelters that the Indians used, as described by Clark, were lodges and huts that resembled each other, but of varying size. They were built in a circular form from 8 to 14 feet high and were supported by four large pillars set in a square at the center of the hut. Around the hut, forks about four feet high crossed with beams stretched from one end to the other to form a circle and support the top. On top of those beams, round poles stretched from beam to beam covered with willow and sod to cover all except the door, which was about 5 feet wide and covered with grass and dirt. A fire pit in the middle provided a heat source and a place to cook. Smoke escaped through a whole in the roof. Some of the lodges housed three to four families along with their dogs and horses. **(1)** These lodges kept out the wind and the cold using the sod and dirt as the insulator.

The Fort that the Corps built was made of poles and timber with little insulation. The spaces between the logs chinked with mud to seal out the wind. The rooms were small and probably stayed warm from the heat of the fireplace. On the 13th of November, Clark moved into a hut; probably not totally completed and sealed or having a heat source, but it was a way to get out of the weather as the days were getting colder and snow was beginning to fall. Ice was beginning to set up on the river and there was still no sign of the hunters. The carpenters continued to work on the fort.

On November 19, the hunters finally returned with a good supply of fresh meat. They had collected 32 deer, 12 elk and a buffalo. Private Whitehouse said in his journal that the total weight was about 2,000 pounds as near as they could tell. **(2)**

A pirogue traveled up the river to collect stone for fireplace chimneys. On the 22nd Sergeant Pryor and five men were sent to the Indian village to get one hundred bushels of corn or about two to two and a half tons of corn that had been given to them. However, they only returned with maybe eighty bushels or about one and a half to two tons.

By the 25th of November, the huts were completed and the men moved into the fort. The temperatures continued to drop and a heavy snow fell on the 27th with wind whipping it into drifts. The daytime high temperatures are now only reaching the teens. The men have collected some food for the winter, but not nearly enough to last the whole winter. They are now in a protected shelter and continue to prepare for the cold temperatures that are coming.

Early December temperatures became rather mild, reaching near the 30s, but as the month went along, they were to experience bitter cold temperatures. During the whole month of December, the mercury only exceeded freezing two afternoons. On seventh of December, a herd of buffalo came within a few miles of the fort. Lewis and party killed 14, three within view of the Fort.

Along with the buffalo came the wolves. The wolves devoured any animal that remained overnight. Lewis was only able to save five of the fourteen buffalo killed. The temperature began to drop to around 1 degree below zero. Clark followed the herd eight miles the next day and killed eight buffalo and one deer. The temperature had fallen to eight degrees below zero. The men were experiencing frostbite so they only returned with meat of two animals. On the ninth, Lewis killed nine buffalo. He remained at the kill site all night to try to save as much of the meat as possible. The buffalo were in such poor condition, only a small amount was edible and the wolves got the rest.

The temperature continues to drop; during the next four days, the mercury only got to the low teens for a high temperature with the nighttime temperatures of twenty-five to thirty-five degrees below zero. On the morning of the 10th, Lewis returned with four horse loads of meat. The Indians predicted that the temperatures would drop more and a storm was coming. Lewis sent out three horses for the remaining meat and instructed the men to return to camp. The temperature on the night of the 12th hit thirty-eight degrees below zero. The temperature came up a bit and on the 14th, it was only twenty below zero. Joseph Field killed a buffalo cow and calf one mile from the camp.

Most of the buffalo now left the river area and hunting was hard. Many days went by with only a few deer harvested. The men hunted as far as 18 miles away from camp without results.

The normal load for packing a horse is around 150 pounds. Therefore, that when Lewis returned with four horse loads of meat, he probably had about 600 pounds or three days rations. If you had just killed a 1,200-pound buffalo, only half of that buffalo is meat or 600 pounds. If you remove the bone, half would be meat, half would be bone and waste. Therefore, each buffalo would probably only net about 300 pounds of meat or only 2 days rations at best.

The Expedition was primarily eating meat, supplementing it with only a small amount of corn or flour. They are conserving those rations as much as they can. Now one begins to see the seriousness of the buffalo leaving the area. Even though the temperatures were thirty-five degrees below zero, hunting still continued if the animals were within one days ride. On December 18, seven men went to hunt buffalo and returned because the temperatures were to cold to proceed. Clark recorded the temperature that morning at thirty-two degrees below zero.

On the plains, the marvelous Chinook winds from the Southwest brought moderating weather with temperatures climbing into the high teens and twenties. However the wind moved the snow around causing large snow drifts and wind chill conditions that seem just as cold as the below zero readings. The wind chill still made the days too cold to proceed.

Most of the men stayed at the fort and on the 23rd they shared a kettle of Indian pudding made from pumpkin, beans, corn and chokecherries with the pits. This dish was a treat to the Indians and was very palatable to the men.

Christmas day rang out with three volleys from the men's rifles and shots from three cannons. They raised the flag. The Captains gave each man a little Taffeta and they danced into the evening. A few men went out to hunt.

Shields and Willard continued to impress the Indians with their blacksmithing skills. They have been repairing equipment and making tomahawk heads out of iron. They traded their services for corn. The Corps brought with them a small forge with bellows fueled with charcoal. The bellows impressed the Indians. Fresh meat was still very hard to come by and the stored supply was depleting by the day.

Two cannon firings brought in the New Year. A few men went to visit at the Indian village. The villagers danced and were impressed with York's agility for a man of his size. The temperature rose into the twenties on January 3 and eight men went to hunt. They returned with nothing more than a rabbit and a wolf.

The weather again turned cold on the fifth and the men were unable to hunt because of the temperature. The temperature for the next ten days was between forty and fifty degrees below. There were many cases of frostbite among the Indians because they were badly in need of meat and attempted to hunt in the bitterly cold weather. On the ninth Clark took a party and joined an Indian hunting party that had spotted a herd of buffalo close to camp. They killed several buffalo, but Clark does not say in his journals if he returned with any of the meat. The temperature was 21 degrees below zero.

Joseph and Rubin Field manage to kill two elk on the 13th. Large numbers of Indians are now hunting. Sergeant Pryor and five men join the hunt. The temperature is around 30 below zero. Whitehouse froze his feet and could not walk back to the Fort. Hunting was proving fruitless.

The temperatures were still below zero and not much relief was in sight. Some days the temperature at midday would get to the teens, but then at night would drop below zero again. Fresh meat supplies were getting very low. A few inches of snow seemed to fall daily and continued to pile up making hunting and travel very difficult. Now the welcome that was given to them months ago by the Indian Chief saying, "...if we eat you will eat and if we starve you will starve..." sunk in. They would have been more optimistic if the temperatures would have moderated and travel was not so cumbersome. Since the large population of the area was all hunting, the scarcity of game stretched for miles.

To pass the time, on January 29, they made a large fire to make charcoal for the blacksmiths. Repairing Indians' articles proved to be a valuable trading tool for food. On the first and second of February a deer was killed each day. A deer only produced about forty to fifty pounds of meat. The meat would be survival rations.

They probably would not starve on that ration but it was not adequate. They had to be eating a lot of corn meal to be satisfied.

On February 4, the temperature moderated to the teens so Clark and sixteen men headed out to find game. Lewis wrote that their supply of meat was completely exhausted. It was also necessary to begin accumulating meat for the journey ahead if possible. Through the next nine days, Captain Clark and his party began to find game and as they killed it, loaded it on horses and sleighs, and took it to caches that would protect the meat from wolves and the ravens. Clark traveled more than sixty miles on his search for fresh meat.

On 16 February, Lewis proceeded to collect the caches and found one had been stolen. The sleighs were loaded and returned to the fort by the 21st stocking the meat house with deer, buffalo, and elk. Many of these animals were very poor due to the hard winter and produced little meat. This stock of meat estimated at 5,400 pounds was enough for around 20-30 days.

The Indians continued to trade at the fort, but fresh meat was hard to come by. The blacksmiths made battleaxes to trade for corn. Each four-inch piece of iron would trade for about seven to eight gallons of corn. Perhaps they would get a deer a day which is just enough for the remaining men if they also used corn to supplement the meat. It was approaching the end of February and finally the weather was moderating with a few daytime temperatures above freezing.

The Corps of Discovery began preparations for continuing the trip up the Missouri. They made four more canoes from large cottonwood trees. They removed the Pirogues from the river and repaired any damage from the ice. Clark and Lewis were busy copying maps and making plans for their continued trip.

The temperatures were now warming and the Indians began to burn the prairies to encourage the grass to grow. On March 5, the temperature stood at forty degrees above zero. Daily activities included trading, keeping the peace with the Indians and continuing preparations for their journey upriver. On March 9, the dugout canoes were finished.

As the temperature climbed into the forties, Clark set all hands to shelling corn and on the 15th, and sun dried parched meal. He began packing articles in eight packs that he had evenly divided among the boats. By March 29, the river began to break up and the ice began to float away. The Indians gathered on the shore to collect the dead buffalo downriver in the ice; this was a delicacy they would have access to only once a year.

By April 1, the Corps was in the final stages of packing and getting ready to depart. Along with their baggage for the upriver trip, they were packing up sundry articles for Thomas Jefferson. Many skins of animals that were new, specimen of insects, mice and the Minitarra Buffalo robe. They packed articles of the Indians and the robe with a war story painted on the hide. They packed these articles on the keelboat that would head downriver when the main party departed upriver. The wind picked up and held the party at the fort for a couple of more days.

Cooking at the Fort during the winter probably used every method; roasting, broiling, boiling and probably some frying. During the entire winter, the journals do not mention one type of cooking method at the Fort. However, On February 14, Ordway and party were hunting and mentioned broiling the meat for dinner. **(3)** They probably did not use the grease or lard for boiling, preferring to save what they had left for the trip up the Missouri. They did extract a little grease from bone marrow during the winter and they probably dried a little meat for the hunting excursions.

Many of the dried protein foods, flour, corn meal, portable soup and the salt pork was saved for emergencies for the trip up the river in case they encountered game shortages. Most of the meat eaten during the winter months was very lean.

Probably the most common cooking method was roasting. Because roasting takes time and they had a lot of time on their hands, the taste compared to that of boiling in water, was far superior. Soups and stews were on the menu, especially with the dried squash, beans, and corn available. Baking Corn Bread with the Dutch oven utilizing the coals from their fireplaces was a common practice. During that period in history, in most homes, part of the fireplace had a hearth used for that purpose. Clark would have built the fireplaces at Fort Mandan the same way. Since the Expedition did not leave any drawings or specifications for their fireplaces, this is speculation but typical cooking of the time. Puddings and cooked corn meal used for breakfast, used water and dried protein. Dried leaves from aromatic plants made good tea.

The hunters probably roasted small pieces of meat over an open fire when they were away from the Fort. They probably did not carry any cooking utensils with them on a hunting trip in the cold weather. Dried jerky, squash and Pemmican was the ration for hunting trips.

Chapter 6
Mandan to the Marias

April 7th 1804 the large boat was loaded with a collection of boxes for delivered to President Thomas Jefferson. At about the same time as the keel boat headed downriver to St Louis, the two Captains, three sergeants, 23 soldiers, two interpreters, one servant and a Indian woman and her infant son, headed northwest up the Missouri. Lewis figured he had provisions for about four months on board two pirogues and six canoes. He stated in his Journal entry for April 7 that this was one of the happiest days of his life. They ere entering an adventure that no other civilized man has ever entered. **(1)**

An adequate food supply was still a problem as they were not far enough away from the Indian settlements to find abundant animals. Spring is in the making with green grass beginning to show and the temperatures are rising. It is too early to harvest any type of plant so living off the land meant any fresh vegetable was out of the question. However, on April 9, Sacagawea harvested some artichokes. These were delicious boiled. Lewis said she collected them from mice burrows where they were stored by the mice for the winter. The Corps was probably using their packed supplies; either what they had accumulated during the winter or the basic rations they brought from St. Louis. The mosquitoes now begin to appear and make life miserable.

One of the hunters was able to kill a deer on the 11th, which was the first fresh meat they had eaten for days. They ate venison and some dried biscuits. These biscuits and some gunpowder had gotten wet a few days prior when water got into one of the boats, but after drying, were usable.

The geese and ducks were migrating north with some beginning to nest. Several geese were shot and eaten. Cooking fowl must have been done by boiling in some of the grease they had left. Roasting fowl takes a long time, which they did not have. I can only presume they used grease. A whole goose or turkey takes about forty-five to fifty minutes to cook in boiling oil at the proper temperature of about 350-400 degrees. They also were able to shoot a few beaver. Beaver is a very tasty animal and soon became one of their favorite foods. They enjoyed beaver tail, as it was mostly fat. A beaver tail that has had the skin removed by blistering the hide, then dropped in boiling oil in minutes a crispy, fatty and not bad tasting tail emerges. The men really enjoyed the tail and they needed a little more fat in their diets. The tail resembles pork rinds, which can be purchased at the market today. The remainder of the beaver meat probably made a good stew.

On April 12, Lewis noted that they had killed two beaver and had found many prairie onions. They collected some and cooked them, probably in the beaver stew for flavor. Prairie onions are excessively strong and it would only take a few to overwhelm a stew. He commented that they were good. In fact, on April 17, Lewis wrote that the men prefer the flesh of beaver to that of any other animal they were able to take. They especially enjoy the tail and liver. **(2)** By using boiling oil, the men could utilize a variety of game like rabbits, geese, deer, antelope and beaver. The mix of animals and bread can all be fried in the same oil adding to its diversity.

It was on this day that they spotted the first tracks of the white bear. This was the fearsome animal described by the Indians to avoid if possible. The buffalo, elk, deer and antelope were now abundant, but the game did not come close enough to kill undoubtedly because the Indians hunted them in the recent past.

As the Expedition proceeded further up the river, the game became much easier to get close enough to shoot. The flintlock rifles only have a range of about one hundred to one hundred-thirty yards; beyond that distance, they are not accurate enough to be consistent. Assiniboine lodges dotted the river's edge. By the time April 25, arrived the animals were so unafraid of the party's presence that the hunters could get meat whenever they needed it.

The buffalo had young calves and the Corps found that the calves were better eating than the older animals. The winter had taken its toll on the older animals and they had not regained their full weight so they were not very good to eat. Sometimes the men only ate the tongue and the marrowbones, as they were not as tough to eat as the rest of the animal. They were now at the mouth of the Yellowstone River with an abundance of animals in all directions. Animals were very easy to kill but the men only took what was necessary to survive.

On April 29, Lewis, accompanied by Charbonneau, met with the first Grizzly Bear the Expedition saw. He said the size of the bear they shot was only 300 pounds. They collected the meat and made some notes that the bears were fierce, but no match for the rifles they carried. Bear meat is fat and full of oil and is probably a welcome change of diet for the men. They did not write about rendering any of the fat. The young bear may not have had enough fat to try to render. They were now dinning everyday on the meat of their choice since the animals remained easy to take. They appear to take both buffalo and elk. They seem to like goose and beaver; these are greasy meats.

On May 2, Lewis again tells of his liking of the beaver. "....the flesh of the beaver is esteemed a delicy among us: I think the tale a most delicious morsel, when boiled it resembles the flavor the fresh tongues and sounds of the codfish, and is usually sufficiently large to afford a plentiful meal for two men." **(3)** When Lewis talks of boiling the tail, he probably means in oil.

I have personally eaten beaver tail and it does not have the taste of water boiled tongue but it is mild and that maybe what Lewis was referring. He refers to the texture as that of the "sounds of the codfish", which is the airbladder of the codfish. **(4)** When the tail is deep fried (boiled in oil) it puffs up like pork rinds and becomes crisp and airy with a crunchy texture. It is interesting to note that Lewis says that the beaver feeds two men. I have butchered an average beaver and found that there is a disproportionate amount of meat on the animal. It is mostly intestines and head. I would guess if you would debone a beaver, you would receive about one and a half to three pounds of edible meat.

On May 5, Clark and Drouillard killed a very large white bear. Lewis estimated the weight to be around six-hundred pounds. He ordered the meat divided and the bear's oil put into casks. The oil when cold is harder than hogs lard. Rendering the grease was a new source of fat for the Corps. Fat has become one of the more important items for the men as it enabled them to cook their food more than four times faster than roasting or boiling in water. They did not want cooking to delay their progress up the river.

On May 8, they were near the mouth of the Milk River. They described the plains as fertile and absent of trees. However, Sacagawea found large numbers of the "white apple" or breadroot and wild liquorices. The breadroot, *Psoralea esculenta* Pursh **(5)** was one of the more important plants Plains Indians used for food. Lewis described how they used it, "...this root forms a considerable article of food with the Indians of the Missouri, who for these purposes prepare them in several ways. They are esteemed good at all seasons of the year, but are best from the middle of July to the latter end of Autumn when they are sought and gathered by the provident part of the natives for their winter store. When collected they are striped of their rind and strung on small throngs or chords and exposed to the sun or placed in the smoke of their fires to dry; when well dried they will keep for several years, provided they are not permitted to become moist or damp; in this situation they usually pound them between two stones placed on a piece of parchment, until they reduce it to a fine powder thus prepared they thicken their soup with it; sometimes they also boil these dried roots with their meat without breaking them; when green they are generally boiled with their meat, sometimes mashing them or otherwise as they think proper; they also prepare an agreeable dish with them by boiling and mashing them and adding the marrow grease of the buffalo and some buries, until the whole be of the consistency of a hasty pudding; they also eat this root roasted and frequently make hearty meals of it raw without sustaining any inconvenience or injury therefrom."

Lewis was constantly looking for plants on the plains that could possibly be of some economic advantage. He continues, "...the white apple appears to me to be a tasteless insipid food itself though I have no doubt but it is a very healthy and moderately nutritious food. I have no doubt but our epicures would admire this root very much, it would serve them in the ragouts and gravies instead of the truffles Morella." **(6)** He suggested this root be used as a substitute for the mushrooms of Morella, which were added to sauces and gravies by gourmet cooks back home.

The Corps still had a store of flour at this time and probably used flour to thicken their gravies instead of the pounded breadroot. However, the tuber itself would work very well as a supplement to their meat stews as we would use the potato today. The Corps chews on wild liquorices, *Glycyrrhiza lepidota*, whose stems are very sweet. Boiling these roots make a thirst-quenching tea. This early in May the plants have not grown very large and the stems are a lot tenderer than they would be later in the year.

Supper on May 9 was described by Lewis and is probably one of the best descriptions of Charbonos' "boudin blanc" or white pudding. (See Appendix II) This was a sausage. The kidney suet is perhaps the most important ingredient as this fat is the purest non-rendered fat of the buffalo. It collects in round goblets around the kidneys and is pure without rendering. The flavor of this fat is superior to any other fat on the animal's body. By using the flour, salt and pepper and the fat, then boiling the lean finely chopped tenderloin or buffalo hump of the buffalo in water, spread the flavor. It would be much like the present day hamburger. Today hamburger usually uses scraps of meat and fat and is not of the best quality. By boiling the sausage in water it melts the fat, flour and seasoning together and of course cleans, "that which is not edible" from the casing. After it cools, it boiled in grease to get a brown crispness and improve the taste. Imagine cooking a hamburger patty in water; the taste would not be that good. If you deep-fry the partially cooked patty, it would be very crisp and would taste excellent. We do this today in our restaurants with chicken fried steaks etc. They are partially cooked then finished in the deep fat fryer.

The animals had become so prevalent that sometimes they needed to be scattered out of the way with sticks. The men are in heaven, considering the availability of meat. Some of the men were scouting for Indian sign when Bratton came upon a large grizzly bear. Bratton was able to get clear shot and wounded the animal through the lungs then informed Lewis where the bear was located. Upon finding the bear two hours later, it was still alive. They could not believe the strength of this animal. Only shooting it in the brains finally dispatched it. The rendered fat was then stored in kegs. Lewis noted that the bear produced 8 gallons of oil. **(7)** This is enough oil for three messes to cook twice a day for about five days before it would be burnt and unusable. Six of the men killed another bear, when the animal threatened their lives. Eight bullets killed the bear. The ferocity of these animals amazed the men, making them very cautious when traveling on land and careful not get in any areas where they might be surprised by one. They processed more oil and stored it in kegs.

The wind on the river was becoming a problem. Everyday wind gusts nearly capsized a boat and almost sank a pirogue on the 14th. Most of the articles that went overboard were retrieved except for a few cooking items. No one described which articles went overboard.

They killed more bear and their fat rendered whenever they could find them. The grease was an important item so every bear the Corps killed they rendered the grease. When the men's clothing started to wear out, brain tanning deer and elk hide replaced the cloth. The trapping or shooting of beaver took place almost daily, for the hides and the food. No one complained being hungry or the supplies running low.

On May 22, they had just passed the Musselshell River when Lewis noted that the game appeared to be less prevalent than the area below. Bear are still around and are shot and rendered, about one every other day. Fishing was almost non-existent. The men caught a catfish that weighed about five pounds, which was the first mention of fish for months.

The terrain was beginning to change. More pines were prevalent and vegetation was less. It looked like the area was very dry with little rainfall throughout the year. The creeks they were passing were mostly dry. The drainages along the river became higher and bighorn sheep appeared. The Captains were excited about the horns of the animals and the use the Indians made of them, but there was not a single mention if they used the animal's meat for food. On May 26, they first sighted the snow-covered peaks of the Rocky Mountains.

At the end of the month of May they begin to see a change from the dry desert that they had just past through to a new more fertile and moist area. The area they were entering was the White Cliffs of the Missouri. Today this area is probably as close to the way it was 200 years ago as anywhere along the trail. Lewis remarked how Mother Nature created the grandest of structures and "...attempted here to rival the human art of masonry". **(8)** Spring rains began to fall and made the travel in the heavy clay soil very difficult. The land was very slippery and when it dried even slightly, mud caked the men's feet in great quantities. They were able to kill enough meat each day since the animals were still plentiful. Lewis thought that they were getting closer to the Great Falls of the Missouri and began to collect elk hides for covering the iron boat. They killed a bear, and of course, rendering of the oil took place. They now were camped across the river from the entrance of another large river and the question was which one was the Missouri.

Strips of meat hung near the campfire. A good example of "jerking meat"

Chapter 7
Marias River to Canoe Camp

The next day, June 3, 1805, the Corps moved camp to an area between the two rivers. It was very difficult for them to determine which river was the Missouri. An incorrect decision would put the expedition behind. They would have to return to the confluence to start up the other river. The Captains decided to camp here a few days and explore the two rivers to determine which one was the Missouri. Lewis was to take the North Fork and Clark the South Fork. About a third of the party remained in camp and busied them selves making clothing and moccasins from hides, and drying as much meat as possible to prepare them for the days ahead. They were also collecting elk hides to cover the Iron boat, experiment they would assemble when they reached the Great Falls of the Missouri.

Lewis and his party moved up the North Fork. There were not many animals within their range and when Drouillard killed a deer, they built a fire and roasted it. They probably used green branches and pierced pieces of meat at the end. Then they held them over a fire to cook. The next day Lewis shot a "burrowing squirrel" and said the meat was flavorful and tender. These squirrels could have been Richardson ground squirrels or a prairie dog. However, Lewis has previously referred to prairie dogs as barking ground squirrels. They used sticks to roast the squirrels.

Clark's party moved up the South Fork and had much better hunting. They killed several bear and dined on marrowbones. The marrowbones are probably leg bone roasts put next to the fire to cook or hung by the bone over the fire to roast the meat. The size of the bone used is proportional the length of cooking time. As the size of the roast increases in weight, the longer the roasting time. The marrow itself is delicious. After roasting, the bone is set next to the fire to finish cooking the marrow inside the bone. It was then cracked open and the marrow eaten. The taste of the marrow is a lot like butter. This method is simpler than roasting with sticks as one does not need to attend to the roasting process as carefully and is less time consuming. The risk of losing the meat in the fire is increased by using sticks.

Lewis' party traveled much further up the North Fork and returned to camp two days after Clark had arrived. Clark was worried about them having troubles because of continuous rain, but on June 8, Lewis returned very wet having experienced some of the most horrible, slippery ground encountered on the journey. A decision was required on which branch to take. When all the party voted, the men went along with the Captains decision to take the South Fork. Lewis named the North Fork Maria's River after his cousin.

Lewis decided the next day to cache some of the articles, possibly to lighten the load as they felt they were approaching the falls of the Missouri. Included in several caches were provisions for the return trip and other items they could do without until they returned and recovered them. On June 9, Lewis noted, "we now selected the articles to be deposited in this cash which consisted of 2 best falling axes, one auger, a set of plains, some files, blacksmiths bellows and hammers, stake tongs &c. 1 keg of flour, 2 kegs of parched meal, 2 kegs of pork, 1 keg of salt, some chisels, a coopers howl, some tin cups, 2 musquets, 3 brown bear skins, beaver skins, horns of the big horned anamal, a part of the men's clothing and all their superfluous baggage of every description, and beaver traps." **(1)**

Captain Lewis must have felt at this point that they were going to have plenty to eat as he deposited a good ration of flour, corn meal, pork and salt. During the winter at Mandan it shows that Lewis was successful in reserving the dried supplies for the trip ahead of them. Another view might be he was considering the return trip and was keeping enough rations for the return.

On June 11, Lewis headed up the South Fork overland to avoid the steep ravines. Because Lewis was sick, they only traveled nine miles before stopping for the day. Lewis, who trained in herbal medicine by his mother, made an herbal remedy from chokecherries that cured him. The next morning they set out very early. After a few hours of walking across the prairies, Lewis decided they needed breakfast and headed back to the river. Upon arrival there, two bears were alarmed and both shot. They butchered both bears and in two hours had roasted and eaten their fill. They tied the remainder of the meat to a tree out of the reach of the wolves, so Clark and the main party passing the area could utilize the meat. Lewis again took to the plains to avoid the breaks of the river and by late afternoon had traveled 27 miles. They rested in the evening and ate heartily. He did not mention how they prepared the meat. In the evening, they went fishing and caught several saugers on deer spleen bait that Goodrich had brought. **(2)**

In the morning, the men ate breakfast before leaving camp, dining on venison and the sauger that were caught the night before. There was not a note regarding how they cooked the fish or ate the remains of roasted fish and venison from the night before. Since they left at Sunrise, my guess is they ate the cold leftovers, which was the common practice in order to progress faster. Lewis hiked 15 miles and at 1:00 pm found the Great Falls of the Missouri. This was a defining moment for Lewis since he now knew they were on the right river, the Missouri.

Later that afternoon he walked downriver to find a location to land the boats. Having found none, he returned to camp where he had one man prepare a scaffold and collect wood to dry meat. The wolves were always present and if any supplies left unattended, the wolves would have them.

The next day Lewis sent a letter back to Clark telling him about his discovery and watch for a place to make a camp before the falls. Then he proceeded up the river and found another 50-foot fall of water. He tried to sketch a picture of the beauty but finally gave up and described it with words as beautiful as the falls itself. Lewis continued westward and to his astonishment found three more falls and thousands of buffalo in the area. He thought he would kill one to have meat if he had to stay the night. After he shot the buffalo, he failed to reload his gun and escaped into the river to avoid a grizzly bear. His only defense was the espontoon that he carried with him. The bear retreated and ran for three miles in alarm. Lewis had learned a great lesson, it is necessary to reload the weapon immediately after every discharge. After seeing another animal he described as a "tyger cat" he decided to return to camp as he found it to be far too dangerous to be in this area alone with thousands of buffalo and many other dangerous animals.

Realizing that the task of portaging around the falls would be a greater task than first imagined, Lewis now instructs his men to take advantage of the abundance of meat available and dry as much meat as they can. He probably also ordered the rendering of as much fat as they could collect. They were in heaven as far as their food supply and they could eat as much as they wanted every day. The vast herd of buffalo being at the same location as the Corps was a coincidence, as the herds were migrating north. There is a hard bottom crossing a quarter mile north of the entrance of the Medicine River. The buffalo used this part of the river to cross on their migration to the fertile prairies to the north.

On June 15, Lewis relaxed and amused himself fishing. He caught several cutthroat trout and had Goodrich dry them for future use. Drying these fish by filleting the fish and hanging the fillet over the tripod the same way they dried buffalo jerky. When drying a whole fish the result may not be as thorough, especially if the fish were larger than two pounds. Filleting the fish and leaving the skin on allows for a very nicely finished dried product much like our salmon fillets of today. Clark continued up the Missouri with the pirogue and dugouts. Downstream from the Great Falls he came to a series of rapids that were not navigable and proceeded to make camp and wait for Lewis to return. Clark was concerned as Sacagawea had been very ill for the past several days and was not getting better.

June 16 Lewis moved downstream to Clark taking with him six hundred pounds of dried buffalo and several dozen dried trout. It would take longer than two days to dry that much meat in the sun. They built fires under the scaffold holding the meat to speed the drying process.

This brings up the question of what is the difference between dried meat and jerked meat. The process of drying meat is just as it says. It removes the moisture in the meat by laying it out in the hot sun or hanging it over a fire. By removing only enough moisture to prevent the blood from dripping from the meat, the procedure slows the spoiling of the meat a few days. During this period, they cut meat into longer slabs about one inch thick.

Slicing jerked meat into much shorter and thinner strips, the drying process takes less time. It might keep for a month. When achieving a 95% moisture removal, the same volume of jerked meat takes considerably longer to dry. This I the jerky we find in many stores today. The dried meat that Lewis transported to the camp could then be cut into smaller pieces and boiled in oil or be made into stews. (See photo on page 66 for size of jerked meat)

One has to remember that the daytime temperatures were reaching into the seventy-degree range and that would spoil fresh killed meat in one to two days while the dried meat would probably spoil in four to six days. The men for their day trips did not carry dried meats, as they were probably still raw. Day trip meats were probably jerked to a much drier product (or cooked meat left over from the previous night's supper) and would be carried in their possible bag without mess.

Now, Clark must consider the problem of getting around the series of falls. He designed a plan to map a portage around the falls now that they fully understood the layout of the land. Lewis stayed behind in the lower portage camp with the main party building carts to transport the canoes and the baggage over the prairie to a point eighteen miles to the southwest of their location. They continued to dry meat, render fat, and made a cache while at this location waiting until Clark returned from marking a portage route to an upper camp.

Sacagawea had been very ill for the past few days, but with treatment of mineral water from a nearby sulfur spring and pain medicine from Lewis's medicine kit she began to feel much better. Sacagawea walked out of the camp and collected "white apples". (Described on May 8th by Lewis *Psoralea esculenta* Pursh, breadroot) Together with eating the roots raw and some dried fish, she again took ill. Lewis treated her and she began to improve. Lewis divided the duties of hunting, fixing the carts for the portage and assisting Clark in measuring the portage route. Not much had been said about food other than it was very plentiful and they were drying meat and fish in order to make it as easy as possible during the portage; it would not be necessary to hunt during that period as all hands needed to be available to transport the baggage and equipment over the rolling plains. This would be the easiest location for obtaining food on their entire journey.

Drouillard, R. Fields and Shannon hunted on the North shore of the Missouri for elk in order to secure enough hides for the iron frame boat. There seemed to be more timber on that side of the river and might make it easier to acquire those hides. Lewis was worried they may not obtain sufficient number to wrap the iron frame.

On the Morning of June 21, the portage began. The first load of goods included the iron boat frame and most of Lewis's gear. The first order of business upon arrival was the construction of the iron boat. This boat, known as the experiment, was to haul most of the baggage, meat, and gear the remainder of the journey. Lewis planned the experiment from the beginning. Its importance to the success of the expedition relied on its completion. The caching of the pirogues below the falls left little choice if it failed.

The hunters on the north shore continued to kill game for the meat and the hides. When Lewis and the men arrived at White Bear Island late in the day they planned to dine on meat left by Clark two days prior. Unfortunately, the wolves raided and consumed the meat. By hanging raw meat for later use, it increased the number of bears and wolves as it attracted them to the smell. The journals provided weak descriptions of the cooking equipment transported of the first load to the upper camp. We can only speculate that at least one brass kettle was on that load. The men were horribly tired from the portage, including the normal cooks so they probably were not interested in spending much time cooking. They undoubtedly boiled the meat and ate as fast as they could so they could get some sleep.

The next morning two of the men looked for wood for the iron boat. The rest of the men returned to the lower camp with the carts to return with another load of equipment and stores. Drouillard and Fields returned to the upper camp and reported that Shannon had not been seen for a couple days. He left their camp and took with him a small kettle, and some parched meal. It was the only kettle for the three men. Its absence of course upset Fields and Drouillard. This gives us some insight as to the method of cooking that the hunters used while away from the base camp.

They obviously boiled or fried the meat in the kettle. They used the parched meal to coat the meat, and then fry it. They used buffalo kidney tallow or other buffalo fat for grease in the frying process. Later in the day, Shannon had returned and processed over 600 pounds of dried meat. Drouillard and Fields collected around eight hundred pounds of meat and hundred pounds of tallow. They transported the hides, meat and tallow to upper camp by canoe.

Capt Clark reported from the lower camp that he had dysentery and assigned Charbonneau to cook for the returning men. Nothing was said what was cooked so it was probably the same kettle arrangement with rendered grease. It is fast and can accommodate a large number of people in a short time. Clark did note however, that he had a little coffee in the morning and that he had not had tasted any since last winter.

The crew was up for an early start to the upper camp. When they reached Willow Run around noon Whitehouse reported that they dined on dumplings and broiled meat. Dumplings usually refer to steamed or boiled bread. However, they probably broiled it over a fire or small strips of meat roasted on sticks. This would enable them to roast it quickly. A boiled dumpling takes about 10 to 15 minutes depending on the size. If they were to steam the dumplings, they would have needed a tight fitting lid such as the Dutch oven. There were no remarks mentioned of the Dutch oven however, Lewis does mention having dumplings with a buffalo stew on the Fourth of July. Placing the dumplings on top of the meat and gravy and then with a tight fitting lid are steamed, they make a delicious stew. It is possible they used the Dutch oven, as they were probably transporting it to the upper camp. A Dutch oven with a little water in the bottom could steam dumplings very quickly. The Corps cached several items at the upper camp but the Dutch oven was not on the list.

Charbonneau rendered tallow and filled three kegs of grease. Work continued on the iron boat at the upper camp. The construction of the boat was beginning to look doubtful since sufficient supplies could not be located. There was not enough elk hide, bark and construction wood. Pitch was also not available since there were very few pine trees in the area.

Lewis was the cook at the upper camp as all of the men were at work on the boat. For dinner, he made a stew with a large quantity of dried buffalo meat and made each man a suet dumpling as a treat. Not knowing that the men probably used a little of the flour they were transferring to make dumplings for lunch, Lewis was going to treat them to a dumpling. That indicates to me the ration of flour was doubled for this day. This is an indication Lewis had a Dutch oven. Placing the dumplings on top of the meat and gravy and then with a tight fitting lid are steamed, they make a delicious stew. (See photo by Chapter 16) However, the men cooked the largest percentage of their meat in oil and in the same oil added flattened flour dough balls. This would make a fry bread. When working as hard as these men were, they would need as much fat as they could eat making the deep fried bread was a treat.

On June 27, Ordway and three other men made a trip to the big spring. Ordway commented as it was the purest, coldest water that he had ever drank. As they approached the springs, the party killed a fat buffalo and took its hump. They broiled the meat over a fire and had a great feast. The buffalo hump is one of the fattest pieces of meat on the buffalo. Marbled fat throughout the roast gives it great flavor. It was a relaxing day for the men and they seemed to enjoy themselves.

The portage struggle continued, but progress was slow as the weather turned violent with afternoon thunderstorms with heavy rain showers. Clark, Charbonneau, Sacagawea and York took a side trip to the springs since the prairies were too wet for travel. A violent rain and hailstorm developed just after they left Willow Run. Some of Clark's equipment was lost but the party survived with their lives. Clark recorded some of the hailstones as being seven inches in diameter. Hailstones pelted the men at camp and several were injured. No one will forget the experience of the punishment dealt by Mother Nature.

On July 2, the portage was completed. White bears were more troublesome every day and the mosquitoes were constantly hampering their labors. Most of the men were putting the final changes to the iron boat, but the success of it looked unfortunate as the parts that were needed, cross bars, bark and tar were in short supply. Daily the hunters supplied enough meat for the camp and collected more each day to dry for the trip upstream. The grizzly bears attracted to the camp by the scent of raw meat became more troublesome. . Whitehouse reported that a bear hunt across the river from the camp supplied a hide and grease. Grease rendering probably was continuing daily to supply cooking grease for the trip ahead. On July 3, the men continued putting the finishing touches on the iron boat. They made Pemmican so that they would have food for the long journey ahead. The Indians at Mandan had warned them the buffalo would become scarce as they approached the mountains. The boat was finished and turned to the sun to dry.

Drinking the last of their spirits, the men danced and celebrated Independence Day. Lewis had prepared a dinner of bacon, beans, suet dumplings and buffalo. He said, "in short we had no just cause to covet the sumptuous feasts of our countrymen on this day." **(3)** The dinner was not a quick one as many of their meals were.

It sounds like Lewis took time on this dinner as a reward for finishing the boat and in celebration of Independence Day. Bacon and beans take time to complete. The beans need to be soaked overnight to break the hard shell around the bean and allow it to be easily cooked. Bacon and salt added to the beans for flavor. Simmering the beans over low heat for hours brings the flavor and tenderness of the beans. A metal spit utilized to hold a large roast over the fire, as the beans were cooking, may have been one of the occasions that buffalo was roasted over a spit. Roasted meat was a treat as the taste exceeds that of other methods. They had the time to tend a good cooking fire throughout the day. A good roasting fire would be a small fire that would hold a temperature around 350-400 degrees around the meat. Rotating the metal spit is necessary to cook the roast on all sides and keep the cooking process even. The spit could have been a piece of blacksmith metal. There was not a description of a cooking spit in the journals but with blacksmith metal available, the construction of one would be simple. If metal was not available, a green branch that would be sufficient to hold the poundage of the meat would be used. In order to serve the number in the party, there could have been as much as one hundred pounds of roast over the fire.

The best roast on the buffalo would be the hump. Bread in the form of dumplings was an easy fix; water, flour and salt mixed together to form a gooey dough ball. This ball is spooned into a Dutch oven with half an inch of boiling water. When spooned into a Dutch oven, the dough ball retains its size and does not boil away but with a tight fitting lid would be steamed and be light and fluffy. It is hard to determine if the use of a Dutch oven occurred, but with this type of cooking, I would think that at least one of their possessions during this period would be the Dutch oven. The Dutch oven lid is very heavy and when in place acts as a sort of pressure cooker and when coals are place on the lid makes a beautiful oven. I believe this tool would be present as long as the burden of carrying the oven would not present itself in such a way that the weight of it would be prohibitive. The statement that Lewis used saying his compatriots did not have anything on them indicates to me that this was a special meal. They danced and were joyous, joking with each other and had a good time on Independence Day.

During the next couple of days, Lewis tried to get the boat to dry so as he could put tar and charcoal on her to seal needle holes and cracks. The men were busy hunting and preparing their personal clothing. They had retained brains from the animals to tan a few deer and elk hides for clothing. On July 7, Sgt Gass quoted in his journals, "the hunters had not good luck, the buffalo being mostly on the plains." **(4)** This was the first indication that the buffalo are at the end of the migration north and most have crossed the hard bottom crossing on the river only a couple miles north of their camp. It was an astounding coincidence that the migration had coincided with the Corps time around the Great Falls of the Missouri providing the most abundant food supply they would encounter on the entire journey. It was good that the abundance existed as it would have taken more hands for hunting and would have prolonged the building of the iron boat even longer.

Lewis now is beginning to have more doubts about his boat floating and is worried the time is taking to long to get her done. He figured she would hold eight thousand pounds of cargo. On July 9, the boat placed in the water floated for a few minutes then started to leak when a windstorm pushed her around. Lewis ordered the boat sunk and to salvage a much a possible. Clark and ten men headed up stream to a large cottonwood area to begin making some dugouts. While Clark and crew were building two more dugouts Ordway and seven men with four canoes began to carry baggage to the area where Clark was making dugouts. It was twenty-three miles by river and eight miles by land to the dugout preparation area. The wind blew hard and made the canoeing upstream very difficult. Men at both camps were busy drying fresh meat for the trip up the river.

On the morning of July 13, Lewis left the White Bear Camp with the rest of the baggage. Lewis walked by land while Ordway and men paddled the canoes with the baggage. When Lewis arrived at the Canoe Camp, all were busy. Some were drying meat by means of a scaffold and fire and others were hollowing the two dugouts. Lewis noted "we eat an emensity of meat; it requires 4 deer, an Elk and a deer, or one buffaloe, to supply us plentifully 24 hours." **(5)** The men now were only eating meat and saving their dried protein for the mountains. Clark had the meat of three buffalo dried and made into pemmican. The pemmican probably did not have berries or fat added, but was pure meat dried to almost cooked state. They may have added tallow but I would guess that probably was not the case. Time did not allow them to pound or add berries to any of the meat. This meat probably had a shelf life of about five to six days before it would turn sour.

They finally have totally finished the portage around the Great Falls of the Missouri and proceeded up the Missouri. They now have eight dugout canoes, which are heavily loaded with meat and goods. Lewis walked by land and easily stayed well ahead of the canoes making slow progress along the meandering river. Fresh meat was still available and utilized instead of the dried stores they had accumulated.

The Great Falls of the Missouri was the first and biggest of five prairie waterfalls Lewis found. Portaging these falls put the Expedition about six weeks behind schedule and added to the almost fatal struggle through the "terrible Rockies" where the Corps of Discovery's ability to survive was severely tested.

Chapter 8
Canoe Camp to Beaverhead Rock

The Captains are moving at a speed that will get them to their destination before winter. July 17th was Lewis' first journal entry that indicated game was getting scarce. There was no word of killing fresh meat. However, there were plentiful fruits that were ripe and eaten as they were collected. The yellow current was in large numbers. The Corps were working hard and had little time to rest. The journalists described no cooking methods. They probably did not make the fruits into any desert with flour, but ate them raw. Dried meat is still available and on board and they were using it. Clark began walking on land to search for the Shoshone Indians and the canoes continued to move up stream. On July 18, Clark killed an elk and left the meat for Lewis and the men. However, it was not enough to feed everyone in Lewis's camp so he issued some of the dried meat. The canyons were deep and seemed as though they were at the gates of the mountains. Sheep were running along the edges of the canyons, but were at a distance to far to be harvested for meat.

After traveling four days up the river, the Corps found themselves in a beautiful valley where the river stretched out to be a mile wide. They were able to kill a couple of swan, a few young geese and a couple of deer. Food was still scarce, but no mention of tapping the dried protein. The fresh meat is still in short supply to feed the entire party to their fill. The dried meat reserves must be getting low and perhaps are getting close to spoiling after being in the hot sun for nearly 10 days. The partially dried meat undoubtedly spoiled as it probably still had some blood in the meat. The meat they dried at canoe camp and made into pemmican is probably beginning to spoil.

Lewis made his way by land while the canoes were negotiating several islands and channels along the river. He discovered a small plain that contained large quantities of onion, which were white, crisp and well flavored. He gathered about a ½ bushel of the onions and when the men arrived in the canoes, they had breakfast; probably some of the pemmican boiled with the onions. The men collected more before they departed.

The first entry indicating that they were not getting their fill of food was from Clark on July 22. He wrote, "having nothing to eat but venison and currents, I find myself much weaker". Later that day the hunters were successful in finding a couple of deer and an elk. Clark was ahead of the canoes a few miles so did not have access to dried meat. His party of men relied on the hunting of the day. If they killed more than they needed they moved it to the river so Lewis and the canoes could retrieve it for their use.

The hunters found it easier to secure meat the next few days although the buffalo was absent since they entered the mountains. The most prevalent animals are antelope and a few deer. Occasionally they spot elk. Clark was still looking for the Shoshone Indians and Sacagawea insisted that she recognizes this country and that they were not far from the three forks of the Missouri. There is Indian sign everywhere but spotting them was not successful.

The Corps encamped at the three forks of the Missouri on July 27, 1805. They needed rest as they had been pushing hard. Clark explored the Southwest Fork (Jefferson River) and the Middle Fork (Madison River). They named the Southeast fork the Gallatin River. Though thoroughly fatigued the hunters were able to find ample game to supply the camp. I would guess now dried meat was not available for backup subsistence. They may have perhaps dried a little fresh meat while they were encamped at the Three Forks of the Missouri. The most important duty during the encampment for three days was to fix new clothing from the collected hides and to rest. Some extra game was collected by the hunters; almost all of it was deer.

They left the encampment on July 30. Lewis took to the shore while Clark and the men canoed up the Jefferson. Lewis was displaced come sundown and spent the night along the river after shooting a duck for dinner. He probably roasted the duck, as he had not planned on spending overnight so had no other means to cook it. The hunters killed only one deer during the day. That probably was not enough to feed everyone. On the 31st Lewis writes in his journals, "nothing killed today and our fresh meat is out; when we have a plenty of fresh meat I find it impossible to make the men take any care of it, or use it with the least frugality, tho' I expect that necessity will shortly teach them this art." **(1)** About noon the next day, Capt Clark killed a mountain sheep. Not having eaten the crew wasted no time to get a fire going and cooked the sheep.

It was Clarks' birthday so he celebrated by issuing a little flour to the crew, which made them a very satisfying meal. Clarks' men were carrying the kettles so this meal may have cooked by boiling in oil since they were in a hurry to eat. Having flour, they may have added water and a little salt and made fry bread, which cooked right along with the meat. Thin slices cut in strips take 3 to 4 minutes to cook and about 4 minutes for the bread. The party was again moving up the river by 3:00 pm. This was the first recorded use of use of flour for weeks. Lewis and his party walking ahead also found game and filled their stomachs. They left the remaining animals along the river for Clark to collect because he was worried about them not finding food.

Lewis went ahead of Clark and the Canoes. He was looking for Indians and a route to bring the canoes. The Jefferson River is dropping in depth and it is becoming very hard to pull the heavy dugouts over shallow rapids. Fresh meat is available but not in abundant supply. On August 5, the canoes were nearly impossible to get upstream. They are now only a mile up the Wisdom River (Big Hole). Whitehouse writes "We came 8 miles this day, & encamped; our party are much fatigued, & it is the wish of all of them, that we would proceed on our Voyage by land to the Columbia River." **(2)** The next day, Drouillard overtook them and he brought news that they were on the wrong fork of the river. As they floated back down to the entrance of the middle fork, the rapids turned over one of the canoes causing them to stop and make camp to dry the goods. Whitehouse lost some of his equipment. The hunters retrieved three deer and one yearling elk. The next morning they unloaded one canoe and left her in the bushes. They proceeded up the middle fork.

Hunting was becoming less successful and the game was not fulfilling to the men. Whitehouse wrote on August 10, "We now have nothing to live on but fresh meat & that poor venison & goats flesh, and our men seem much fataigued; and find that meat only, is too weak a diet, for men undergoing so much fataigue; but they seem all content with what we can get." **(3)** Clark and the dugouts passed the Indian landmark known as the Beaverhead Rock; a hill that resembles the head of a beaver. They were now sure they would find the Shoshone soon.

Chapter 9
Beaverhead Rock to the Nez Perce

On August 11, Lewis was several miles ahead of Clark and the canoes. About mid morning, he finally saw a lone Indian rider. They tried to make contact with the man, but frightened him away and lost track of him as he rode over the hills. Clark and the dugouts proceeded to a large Island that he called 3,000-mile Island, the distance he estimated they were from the mouth of the Missouri. Game was barely enough to supply his crew with minimal subsistence. They recorded the abundance of beaver in the valley for future trapping. Whitehouse wrote, "The beaver at this place, are more plenty, than at any place we have been at, since we entered the Mesouri River". **(1)**

Lewis, McNeil and Drouillard were following an Indian trail that appears to go over the mountains. They stopped for breakfast where the stream turned to the north. Lewis entered the following, "here we halted and breakfasted on the last of our venison, having yet a small piece of pork in reserve". **(2)** The river turned into a creek and Lewis wrote, "...McNeal had exultingly stood with a foot on each side of this little rivulet and thanked his god that he had lived to bestride the mighty & heretofore deemed endless Missouri." **(3)** A little later in the day, Lewis reached the top of the continental divide and was stunned to see endless mountains to the west instead of the Columbia River he had expected to see. They camped near the top of the continental divide for the evening. Since they had not acquired any game during the day supper was the last of their pork and a little flour and parched meal. There were still berries available, but they were very bitter.

It was now August 13; Lewis proceeded down the Indian trail and surprised a party of four Indians, three women and one man. Lewis approached them with peace and convinced them to take him to their encampment. Upon arrival, they smoked and ate pemmican made with dried berries. The Indians complained as this was the only food they had. Lewis wrote in his journal. "on my return to my lodge an Indian called me in to his bower and gave me a small morsel of the flesh of an antelope boiled and a piece of a fresh salmon roasted; both which I eat with a very good relish; this was the first salmon I had seen and perfectly convinced me that we were on the waters of the Pacific Ocean." **(4)**

Clark continued up the shallow river. The men were very fatigued and progress was very slow as they pulled the dugouts over rocks. They wanted to quit and begin the journey over land. Food was not as scarce as it was for Lewis. The main food was deer, antelope and trout. It gave the men very little strength because it was very lean. They were probably getting a ration that was equal to about half of what they need. The river they were following became so small that they were barely able to use the canoes. In the next few days, they would need to begin an overland trip.

Lewis miles ahead of Clark and in the encampment of the Shoshone, decided to wait a day at their camp, observe and acquire as much information about the geography as possible. There was no fresh meat in the Indian camp so Lewis sent his hunters out to find food. So did the Indians, as they were hunting on horseback and were unable with a large group to get meat. Neither McNeil nor Drouillard, who were on foot, could find game. Lewis wrote, "I now directed McNeal to make me a little paist with the flour and added some berries to it which I found very pallateable." **(5)** These Indians were very poor and food was scarce. The geography lesson was troublesome. Lewis learned that game was almost nonexistent going over the high, snow-covered mountains. They may have to go hungry for days if they attempted the journey.

Lewis talked the chief of the Shoshone into assisting in retrieving the baggage with his horses. They left the encampment with nothing to eat. Lewis had about two pounds of flour with him that he divided into two parcels; one for the morning and one to be used for the afternoon. They mixed the flour with a little water and added a few berries. That evening when they camped, they ate the other half in the same manner. None of the group of Indians or hunters acquired game that day. Lewis proceeded in Clark's direction, while he sent the hunters out to find food.

Late in the morning, one of the Indians watching the hunters came riding full speed and told them that one of the hunters killed a deer. All of the Indians ran their horses at full speed toward the kill. Lewis' description of this is beyond any other. He wrote, "the fellow was so uneasy that he left me the horse dismounted and ran on foot at full speed, I am confident a mile. When they arrived where the deer was which was in view of me they dismounted and ran in tumbling over each other like a parcel of famished dogs each seizing and tearing away a part of the intestens which had been previously thrown out by Drouillard who killed it; the seen was such when I arrived that had I not have had a pretty keen appetite myself I am confident I should not have taisted any part of the venison shortly. Each one had a peice of some discription and all eating most ravenously. Some were eating the kidnies the melt and liver and the blood runing from the corners of their mouths, others were in a similar situation with the paunch and guts but the exuding substance in this case from their lips was of a different discription. One of the last who attacted my attention particularly had been fortunate in his allotment or reather active in the division, he had provided himself with about nine feet of the small guts one end of which he was chewing on while with his hands he was squezzing the contents out at the other. I really did not untill now think that human nature ever presented itself in a shape so nearly allyed to the brute creation. I viewed these poor starved divils with pity and compassion". **(6)** Clark had not yet arrived at the rendezvous point so they waited until morning. They had killed game enough to satisfy their daily needs, built a fire and roasted the meat..

Clark was still struggling to get the dugouts up the river. Food was scarce. The hunters shot one deer and brought it to camp to feed 30 people. This is about ¼ of a day's rations. The area was so void of wood that willow branches had to be collected to make a fire. Since the fire was small, they probably boiled their venison in oil, as it would take less time. The rendered grease must have been getting low also. Nothing was said about the supplies on board; perhaps they were using them sparingly. They called this area Serviceberry Valley as Sacagawea, Charbonneau and Clark found great quantities of these berries.

Knowing that Clark must be close, Lewis sent Drouillard to find Clark and tell him of their presence. McNeil cooked the rest of their meat for breakfast. Later in the morning, word came back that Clark and crew were on their way. When he arrived, they found that Sacagawea's brother was chief Camawait. The celebration among the Indians was joyous. The Captains met with the Indians and planned the trip to the Columbia. There were two options. The first option was by water down the Salmon River. The other was by way of the mountains on an Indian trail. The Indians warned that either way this time of year game was scarce. Ordway felt that the Captains really did not believe them. The Captains traded some of the goods to the Indians and in return obtained enough horses to assist in the overland journey.

On August 18, Clark selected a party to inspect the water route. He took eleven of the men and many of the Indians. They took axes and tools to make boats. The Indians said it was not a good route but Clark had to see for himself. Food was still scarce; the hunters succeed in killing only one deer to feed the group. At Lewis' camp the men continued making saddles and arranging the packs in preparation for the trip over the mountains. The hunters were unsuccessful in finding game. The group relied on trout they caught by making a fish trap out of willow. They also trapped beaver for subsistence. On August 21, Lewis wrote, "neither of the hunters returned this evening and I was obliged to issue pork and corn." They now are ready for the arrival of the horses to move the baggage overland.

Clark arrived at the Salmon River and saw the Indians using a weir to trap salmon. They were given boiled and dried salmon. He found these Indians were very honest, punctual, and very generous even though they were very poor and plundered by other tribes. The only things they really owned were horses. They were short of shelter and a means to protect themselves from their enemies.

Lewis, waiting for the Indians to return with horses, sent Drouillard out to hunt. He approached three Indians digging roots, one male and two females. After a small conversation, the young male stole Drouillard's gun, left his packs and ran with his horse to get away from him. Drouillard pursued and retrieved his gun.

Upon return, he retrieved the baggage of the Indians and took it to Lewis. The bags contained several species of roots and about a bushel of serviceberries. These Indians, in the search of food, utilized roots, as there were times when game was scarce. The collection of these roots was particularity interesting to Lewis now that they were also finding food hard to find. Lewis tasted the roots, some of them boiled and some raw. He liked vegetables and found these roots quite agreeable. He asked the Indians to point out the plants for future reference, which he described in detail. They were the Bitterroot, Western Spring Beauty, and the Tobacco Root.

The Indians, Charbonneau and Sacagawea arrived late in the morning with horses. A party of about fifty came to help transport the baggage. Without fresh meat, Lewis and the party starved. He boiled corn and beans from his dried protein supply to feed them. He told them about growing squash. Lewis gave the chief some dried squash. Lewis writes, "he appeared much pleased with the information. I gave him a few dried squashes which we had brought from the Mandans; he had them boiled and declared them to be the best thing he had ever tasted except sugar, a small lump of which it seems his sister Sah-cah-gar Wea had given him." **(7)** That evening with no game Lewis instructed the men to make a bush drag and they captured 528 trout. The Indians did not have any food so Lewis shared the fish. Whitehouse described the fish as pan trout, indicating that the size was not that large.

Clark and his men were still exploring the Salmon River to be sure that the river was not passable. He confirmed his finding saying he had not seen an area that remote on the entire voyage. Game was just too scarce to proceed in this direction; starvation would be inevitable. He proceeded back to the camp.

Lewis finally purchased nine horses and a mule from the Indians. They were to be used to pack the Corps' baggage. He had a cache dug and several items placed in the cache. The journals did not describe the contents of the cache, but probably were heavier items. He also pulled the plugs on the canoes and sank them in a pond. They began the two-day trek to the Indians' camp. The hunters were unsuccessful and provisions were running too low to supply the Indians with any more corn and beans. He informed Chief Cameahwait to direct his people to leave the party and return to their camp. The next day the hunters shot three deer. They divided the deer up between the natives and his men. There was not nearly enough for everyone.

The date was August 26, and the nights were getting colder. They stopped at the top of the continental divide near a creek that is the beginning of the Missouri river or the waters heading down the Missouri. Not having fresh meat Lewis noted, "I directed a pint of corn to be given each Indian who was engaged in transporting our baggage and about the same quantity to each of the men which they parched pounded and made into supe." **(8)** Lewis observed some the starving females digging the roots of Gardiner's Yampa and the seeds of the ripe sunflowers. They smashed these root and seeds to make flour.

Clark's camp also had very little to eat. Only a few fish, no game and a little dried salmon from the Indians. Hunger was beginning to affect the men. Clark sent Colter to find Lewis at the Shoshone camp to tell him that the river was not navigable and to buy horses for the trip over the mountain. The warning of the shortage of game on mountain pass and their dried supply of food being very short did not yet seem to concern them.

On August 27, Lewis did not make a journal entry. He is obviously concerned about the overland trek they are about to make. He purchased more horses and began final preparation to move to Clark's camp a few miles below. He did not have enough horses yet, only twenty-five, but decided he may have to make do with what he had. Food is very scarce; both parties are mostly living on fish. Clark writes "Those Sammon which I live on at present are pleasent eateing, not withstanding they weaken me verry fast and my flesh I find is declineing". **(9)** Lewis was not having much luck buying more horses.

On August 30, they decided to proceed on with their journey to the ocean. There is still no game. The Indians began to pack up and head to the Missouri in search of buffalo. The Corps completed a few more packsaddles and was ready to move. The next day the Corps headed up the creek with 30 horses packed and the men walking. They were going up the North Fork of the Salmon River into the mountains and Flathead Indian Country.

The Flatheads and the Shoshone tribes meet and winter on the Missouri, where buffalo where is plentiful. Lewis still is not writing in his journal. There was very little game, only an occasional deer or fish; however, they purchased a little dried salmon from some Indians along the route. They are making about 20 miles a day. The Indian guide that was with Clark up the Salmon River "The River of no Return", remained with the Corps. They named him Old Toby, it being the best translation from the Shoshone to English.

The going was tough. The mountains were so steep that the horses sometimes give out. The distance covered per day was now around twelve to fourteen miles. Game is not available except a few blue grouse, which they may be able to shoot. The dried Salmon is almost gone. It is September 3 and fall is upon them with winter approaching fast, nights are cold, rain turns to snow. The party of thirty had run out of fresh meat except three or four grouse each day. Ordway writes, "eat the last of our pork &.C. Some of the men threaten to kill a colt to eat they being hungry, but puts if off untill tomorrow noon hopeing the hunters will kill Some game." **(10)** Toby continued to look for the trail to the valley and finally descended. The hunters were able to kill grouse and a couple of deer. They had enough food to satisfy them. They proceeded on until they came to a village of Flathead Indians with four to five hundred horses. They had little to eat at that time. Cooking probably took more time now since they were probably out of grease.

The journals did not mention the cooking methods or if the military option of three messes was continued. If they were out of grease, they were probably roasting meat over the fire. They probably made one or two larger fires as the temperature was cold and roasted the meat in small portions for each man. They usually had to give the horses a rest and let them graze, so every time they stopped they spent one and a half to two and a half hours. This was plenty time for each man to roast a one to two pound pieces of meat. They may have boiled the meat in water with the kettles, but that would take as much time as roasting and would not be as satisfying. The Flathead Indians were subsisting on dried serviceberries and roots. The Captains were able to trade seven of the injured and sore horses for eleven good horses. Trading was difficult because everything translated from Salish to Shoshone, to Hidatsa, to French, to English. Then the reply had to go from English to French, to Hidatsa, to Shoshone, to Salish. Sometimes a lot was lost in interpretation.

They continued for the next few days down the Bitterroot River and were able to find elk and deer enough to feed the men adequately to stop their complaints of hunger but not enough to dry for the future. They stopped for a few days to rest the horses and men. Their idea was to head west into the mountains again to find better hunting and easier traveling than when Toby was apparently lost. They called this place Travelers Rest. Lewis again began writing in his journals. Hunting was not that productive; the hunters only shot four deer, a beaver and a few smaller animals. Colter brought three Nez Perces Indians to camp and one agreed to take them to his people located on the Columbia River. They thought it was a six-day trip.

It was September 11, and the days were warm, but the mountains that they were going to cross were filled with snow. They left Travelers rest late in the afternoon. The hunters had no success in securing game. The Corps made about 20 miles the next day over very steep mountains. The hunters were successful in securing four deer; enough meat to reduce hunger slightly but not enough as each man would have liked. The Indian trail was the well traveled Nez Perces Trail.

Game is non-existent and the men are hungry. In want of meat Ordway writes, "had nothing to eat but Some portable Soup. we being hungry for meat as the Soup did not Satisfy we killed a fat colt which eat verry well at this time." **(11)** Portable soup was a military staple and did not taste very good. Extracting the clear sticky material remaining in the roasting pan following the cooking of beef or lamb meat, it is dried and rolled into balls the size of a half dollar.. This would resemble our bouillon cubes today but without salt. There were no vegetables or other ingredients added. The "glue balls" make a base for soups of other products such as roots and dried meats. It may have added flavor but had little nutritional value. The Corps was totally out of grease and proceeded to roast the horsemeat. At this point, very pleasing and not bad tasting. Gass wrote, "without a miracle it was impossible to feed 30 hungry men and upwards, besides some Indians."

The date is 15 September. The Corps breakfasted on the remainder of the cooked horse and continued through the rugged mountains. The hunters were not successful in finding game. The camp that evening was on the top of a mountain with no water. Whitehouse makes a journal entry, "So we Camped on the top ridge of the mountain without finding any water, but found plean[ty] of Snow, which appear to have lain all the year; we melted what we wanted to drink and made or mixd a little portable Soup with Snow water and lay down contented." It began to snow and the temperature was dropping. By evening the next day there was eight inches of snow on the ground. They are high in the mountains but dropped off the mountainside to a valley in the evening to encamp. Clark saw deer that were in range of a shot, but his gun would not go off. Probably from the wet conditions and the flint perhaps was not tight. He tried it seven times before the deer bound away. No game today. Portable soup is the only food they have. That evening they killed another horse. They roasted ½ of it that evening and ate it. Whitehouse writes "The party were all much fataigued & hungry, our officers had a Colt killed and the party eat the half of it this evening. In the evening it quitted snowing, but the wind was very chilly & Cold..." **(12)**

They continued along the steep mountains but progressing downward. The temperature began to rise, but the melting snow made progress slow and wet. There was no meat for the day. However, there was promise of game ahead as they heard wolves howling and saw deer sign. Whitehouse explains, "The party being all exceeding hungry we were obliged to kill a sucking Colt to subsist on. One of our hunters went out hunting. He chased a bear in a Mountain; but did not get a chance to kill it. The Wolves howled very much in the Night, & we saw some signs of deer, so that we expect that their is game to be had a head of where we are encamped". **(13)**

Clark took six men and scouted ahead for food. He climbed a high observation point and could finally see the prairie again. A remarkable sight, I am sure, but it is still miles away. No animals killed and the men again turn to eat portable soup since all of their provisions are gone. Gass makes his journal entry, "Having heard nothing from our hunters, we again supped upon some of our portable soup. The men are becoming lean and debilitated, on account of the scarcity and poor quality of the provisions on which we subsist: our horses' feet are also becoming very sore." **(14)**

On September 20, Lewis continues toward Clark who is several miles ahead. The food is all gone. Ordway writes, "we found a handful or 2 of Indian peas and a little bears oil which we brought with us; we finished the last morcil of it and proceeded on half Starved and very weak; our horses feet gitting Sore." **(15)** In late morning, they found a half of a horse that Clark left for them. They dined on it and were satisfied again, temporarily.

Clark proceeded down the mountains to the plains and found the Nez Perce. The Indians gave them small pieces of dried buffalo, dried berries and camas bread. Clark noticed large quantities of the Camas bulb in small piles drying. In the fall of the year, they were collected in marshy areas for winter use. Clark describes the cooking process. His journal entry reads. "dig a large hole 3 feet deep Cover the bottom with Split wood on the top of which they lay Small Stones of about 3 or 4 Inches thick, a Second layer of Splited wood & Set the whole on fire which heats the Stones, after the fire is extinguished they lay grass & mud mixed on the Stones, on that dry grass which Supports the Pâsh-Shi-co root a thin Coat of the Same grass is laid on the top, a Small fire is kept when necessary in the Center of the kile &c." **(16)** Clark noted that they were rather ill in the evening as they may have eaten too much of the bulb. This bulb is probably the most important winter survival food the Indians used. It was used in about everything they cooked. Near these Indian villages are vast fields of *Camassia quamash.* When these plants are in bloom, they look like a blue sea. The name of this area is Quamash Flats.

Lewis was beginning to get impatient to get out of the mountains. The men were weak and had a lot of bad luck with the horses; they were making poor time. They camped that night and Lewis insisted on an early start in the morning. They still had very little food but Lewis writes on September 21. "we killed a few

Pheasants, and I killed a prarie woolf_ which together with the ballance of our horse beef and some crawfish_ which we obtained in the creek enabled us to make one more hearty meal, not knowing where the next was to be found." **(17)** Again getting a late start against Lewis's wishes, they proceed on down the mountain. In early afternoon, Fields arrives from the Clark party bringing some camas bulbs and a little dried salmon; their hunger was temporarily satisfied. They reached the Indian camp, were welcomed with more bulbs, bread, and dried fish. Lewis and Clark, the Corps of Discovery have now made it to the Pacific side of the Rocky Mountains.

Part of an elk hung in a tree waiting for further use. Several times the journals tell of an advance party or hunters hanging meat in a tree out of the reach of wolves and other predators for the main party to retrieve.

Chapter 10
The Nez Perce to the Pacific Ocean

The Captains were trading trinkets and cloth to the Chiefs in exchange for dried salmon, camas bulbs and bread. One of the very interesting observations was from Whitehouse when he noticed that the natives had cooper kettles. They appeared to be trading with white men from the west. The tribe of Nez Perce dressed in finely tanned hides and had leather teepees.

The Corps of Discovery followed the Clearwater River and camped at a fork in the river. The hunters were able to shoot some deer, which were very agreeable with the men who were all sick from eating too much salmon and camas. It is probable that they had contacted food poisoning. They had stomach cramps and some were vomiting and had very little energy. On Sept 24, Clark writes, "Several men So unwell that they were Compelled to lie on the Side of the road for Some time others obliged to be put on horses._ I gave rushes Pills to the Sick this evening." **(1)** Giving the men Rushes Pills probably made the dysentery worse. Food poisoning from bacteria such as Salmonella usually has an onset from six to forty-eight hours and the symptoms are nausea, fever, chills, vomiting, diarrhea, abdominal cramps, and headache. **(2)** The sickness can persist for one to ten days. **(3)**

Clark went upstream to find a supply of large trees to make dugouts. The next day they moved their camp across the river from the North Fork of the Salmon and prepared to make dugouts from the tall pine Clark had located. Five crews were working on building canoes. Game was still very scarce, but salmon were very plentiful. Gass writes, "Game is very scarce, and our hunters unable to kill any meat. We are therefore obliged to live on fish and roots, that we procure from the natives; and which do not appear a suitable diet for us. Salt also is scarce without which fish is but poor and insipid. Our hunters killed nothing to day." **(4)** Lewis was still very sick. Most of the men were recovering, but they were still very weak. Consequently, they were slow at progressing making the canoes. The hunters were supplying a deer or two each day but this was not sufficient meat for this crew. They continued to live on fresh salmon. As long as they roasted or boiled fresh salmon, they seemed to avoid illness. Clark was giving the men Rushes Pills, along with the camas roots, (which in themselves are a purgative until the person's system adjusts to them) prolonged the gastric discomfort and time to heal from the bacterial infection.

They continued to work on the canoes and on October 2. Since they had not found any game for a couple of days, one of the hunters killed a coyote which they ate that evening. The captains could see the men's hunger and ordered a horse killed for food. The men ate as though this was the best tasting meat they ever had. The Expedition continued buying camas, but the men found that when they ate them they would get gastric pains. Lewis started to feel a little better by Oct 4 and began to walk around a little. Their horsemeat is now gone so a couple of men killed, roasted and ate a fat dog that had followed them from the last Indian camp. They branded 38 horses and left them in the care of the Indians for their pickup when they returned from the coast.

The canoes were tested and loaded; around mid afternoon on October 7, the Expedition proceeded down the river. The rapids were strong and made the passage hazardous. They hit a rapid by the Potlatch River and damaged one of the canoes causing many of the articles to become wet. They camped to dry out and repair the canoe. Game was scarce for days. They bought salmon from the Indians along the shore. Gass writes, "We have some Frenchmen, who prefer dog-flesh to fish; and they here got two or three dogs from the Indians." **(5)** Whitehouse commented in his journal that they purchased a quantity of salmon and a small amount of bear's oil or grease. **(6)** This is interesting as it is the first mention of grease or oil for a month. As they proceeded down the river, they continued to buy dogs and fresh salmon from the natives. Clark did not like dog and found it interesting that some of the men prefer the dog to salmon; but it was not to his taste. The tribes owned several brass or copper kettles, and the men saw a teakettle. They observed various brass trinkets and fishing net. Another civilized nation was obviously trading with these Indians.

They advanced down the Snake River and the landscape turned barren.

Firewood was hard to find. They found that Sacagawea being Indian with the Corps was a sign of peace and made it easier for them to make peace with each group they meet. On October 14, they approached some bad rapids and turned over a canoe losing several articles including one of the small brass kettles and a couple of spoons. That evening they ate eight ducks and thought they were delicious. That was not enough for the entire crew, but a few men got a break from the salmon. The rapids were fierce the next couple of days; ducks and geese supplemented the Corps fish diet. October 16th, the event they had been waiting for happened; they turned their boats into the Columbia River. This was an important trade area for the Indians of the region. Several different tribes were located in this area. The Yakima's the Wanapams, Walula, Umatilla, and Palouses. They all spoke a similar language to the Nez Perces. The Expedition camped near the convergence of the Columbia and later made trades to the Indians for eight dogs, fresh salmon and about 20 pounds of dried horsemeat.

Clark took a few men in a canoe and proceeded north on the Columbia where he saw vast quantities of salmon being dried. The natives built scaffolds covered with mats made of rushes that grew along the river. Filleted salmon, with the skin still intact, and placed facing the sun, were drying on the scaffold. There were very large quantities of salmon that had already been dried laid in storage in piles for the winter. Clark saw that these people survived almost entirely on salmon. Very few horses were present. Their primary mode of transportation was the canoe.

The Indians invited Clarks' party to their camp and fed them boiled salmon. The entry in his journal read. " I was furnished with a mat to Sit on, and one man Set about prepareing me Something to eate, first he brought in a piece of a Drift log of pine and with a wedge of the elks horn, and a malet of Stone curioesly Carved he Split the log into Small pieces and lay'd it open on the fire on which he put round Stones, a woman handed him a basket of water and a large Salmon about half Dried, when the Stones were hot he put them into the basket of water with the fish which was Soon

Suficently boiled for use; it was then taken out put on a platter of rushes neetly made, and Set before me. They boiled a Salmon for each of the men with me". **(7)**

Clark recorded in his journal; the number of dying salmon was so large that he speculated their use for fuel as there was very little wood in the area and it seemed like a logical use. We know today that what Clark saw was the end of the salmon run for the year and salmon normally die after spawning. On his return to camp, he found that his men had traded for salmon and 26 dogs for food. The hunters had shot several prairie chickens and a few ducks.

They continued down the Columbia the next few days and encountered several other Indian villages where they smoked and traded. They noticed signs of white traders as some of the tribes had copper kettles, knives, red cloth and one Indian had a sailor's coat. Wood was still very hard to find for cook fires. There were several comments from the various journals about the scanty dress of the natives; some barely had enough hides to cover the private parts of their bodies. It seems as though every time hunger was at hand and there was not enough fish or fowl available, they ate dog. The dogs were not large animals and probably had only 4 to 8 pounds of edible meat. This is strictly a supplement of their diet as it was not enough to feed the entire Corps.

On October 21, Clark noted Collins had made some beer. "one of our party J. Collins presented us with Some verry good beer made of the *Pa-shi-co-quar-mash* bread, which bread is the remains of what was laid in as [X: *a part of our*] Stores of Provisions, at the first flat heads or Cho-pun-nish Nation at the head of the *Kosskoske* river which by being frequently wet molded & Sowered &c." **(8)** They have not talked of any kind of homemade alcoholic beverage prior to this. I am not sure it was a planned event but Collins evidently understood the fermentation process and helped it along. Clark described it as an accidental wetting and souring or fermentation. There evidently was not a lot of it as it was not mentioned in any other journal.

They proceeded down the Columbia and passed large number of Indians fishing. They had large stockpiles of dried fish and roots. The Expedition traded daily for fish and roots and killed only a few ducks and gulls. Some Indians living near the John Day River used their horses to help the Expedition portage the falls on the Columbia just below the confluence. It was October 22; the Indians told them that white men had been camped in the same spot about a week earlier. They purchased some dogs to eat, as the natives below the falls were not that enthusiastic about selling their fish. The Captains traded one of their smaller dugout canoes for a larger one that was designed to take the rapids and not be as likely to overturn.

They proceeded on to portage a second cascade located at The Dalles. They drifted the canoes down the rapids one at a time. They all made the trip intact. They traded for a supply of dried salmon and procured a little bears grease. This was the second mention of grease on the west side of the continental divide. The grease gave them an added cooking method and taste for their monotonous diet.

On October 26, they decided to layover to hunt and dry some of the equipment. They camped at Mill Creek where it converges with the Columbia River. This is now located at The Dalles. They were visited by Indians that evening and Clark described his food for the day, "our hunters killed five Deer, 4 verry large gray Squirrels, a goose & Pheasent, one man giged a Salmon trout which we had fried in a little Bears oil which a Chief gave us yesterday and I think the finest fish I ever tasted". **(9)** The "salmon trout" was probably a steelhead trout, which is not quite as oily as some salmon. They used a little oil, fried the filleted fish in the bottom of the kettles, and did not deep-fry in the oil. Perhaps if they coated it with the root flour it would have been delicious.

The river from this location is now rising and lowering a few inches everyday, which would be the first sign of the tide, an indication that they are approaching the ocean. They decided to stay another day at this location, as it was the best hunting area they have seen for sometime. The wind on the Columbia can be fierce at times and this may have played in the decision to stay at this location. Gass writes, "This was a fine clear morning, but the wind blew very hard up the river, and we remained here all day."

(10) The next day they visited an Indian camp where they observed the Indians had a British musket, a sword and copper teakettles. They treated these items with care and were very fond of them.

They were now entering an area that is full of pine and deciduous trees so it was very green. The rainfall in this area was substantially more than the territory they just left. Indians of several different tribes reside along the Columbia. The Expedition began to notice waterfalls along the shoreline, which is steep on both sides of the river. The country is beautiful compared to the dessert they just left. They looked forward to the probability of abundant of game. The Corps had purchased 20 dogs over the past few days, as the diet of salmon was growing old. Other game was still hard to hunt because of the severity of the terrain and the speed at which they were moving down stream. Clark spotted a large condor and noted it was the largest bird he had ever seen. They camped above the Cascades and could hear the roar of the falls ahead of them. Upon examination, they found that they would need to portage about 2.5 miles to get around these cascades. They had ample wood now to enjoy the luxury of a fire, as it was raining and chilly. They were probably roasting various types of meat; dog, deer, ducks to eat along with salmon and roots.

On October 31, Clark and two men went ahead to see if the river would allow the canoes to be floated. He left Lewis and the rest of the men to portage the canoes and baggage around the falls. Floating the canoes down the falls was very strenuous work and allowed the men to move only two during the day. They carried the baggage below the falls and reloaded. In the evening one of the men shot a goose. As it was floating down the river, one of the Indians jumped in and retrieved the goose. The area was a particularly dangerous area with swift rapids and large rocks. When he climbed on shore with the goose, he took it to Clark. Clark told the Indian he earned the goose and to keep it. Clark seemed surprised at the hurried cleaning of the goose and wrote, "Suffered him to keep it which he about half picked and Spited it up with the guts in it to roste." **(11)**

They completed the portage around the rapids and falls and camped near today's North Bonneville. More rapids still lay ahead.

Upon Clarks observations, it was decided that it was best to portage another mile before loading. Indians in canoes loaded with dried fish were passing and going to trade with the whites ahead or other tribes. Whitehouse described the passing of one of the canoes. He wrote, "Those Indians had a musket which the stock was made of Brass & Copper & a Powder flask." They were now approaching white traders but there seemed to be little alarm. There seemed to be no questions in the journals as to who these peoples were and where they were from. During the day, some of the hunters had secured 16 geese, brants and ducks, which they roasted for dinner when they made camp near now Rooster Rock.

The fog was heavy in the morning preventing any movement until late morning. Clark noted that Labiech had killed three geese on the fly. This is an astronomical feets as these were flintlocks and shooting a 54 caliber ball. Just hitting a fixed target is difficult, but having it moving is unthinkable. The marksmanship of the hunters was astounding. They had killed a deer in the morning and took it with them for supper. They camped that evening on a large island now known as Government Island just north of Portland, Oregon. The natives told them two ships were two days ahead. Excitement stirred within the troops; the ocean is not far ahead.

There were many Indians in the area and villages all along the river. One Village had fifty canoes on the shoreline. The land was very fertile with an abundant variety of berry plants. The Indians continued presence was beginning to become disturbing, as they were always present with the men. This was good as they could trade for food, but bad because they had no privacy and were always on the watch for thievery. The Indians stole Clark's pipe tomahawk, which he used to smoke with the tribes. They also had stolen one of the men's capote. The Expedition did not trust some of these people, but they had supplied them with four bushels of roots that made fine bread. Gass wrote, "The roots are of a superior quality to any I had before seen: they are called whapto; resemble a potatoe when cooked, and are about as big as a hen egg." **(12)**

The Indians possessed many articles of European origin including muskets and pistols. They took note that one canoe returning possessed a sturgeon. Game seemed to be abundant with elk and deer sign visible along the shoreline. The hunters were now able to get meat and fowl daily. The river was about a mile wide. They camped for the night after making about 30 miles. Clark noted that this was one of the few nights they were absent from Indians. It rained frequently making everything wet and cold; the humidity was high. The men were complaining of fleas they had contacted from the Indian lodges along the rivers. The journalists had commented how fleas infested some of the tribes. Fleas are common today along the pacific region.

They camped under a cliff of rocks on November 6· The men were chilly and, built large fires so they could dry their clothing and possibly kill some of the fleas. They traded for some fresh fish, as they had killed nothing during the day. The brush was very thick and wet so the rocks were a drier location. The next day was still rainy and wet so when they met several Indians they traded for dogs, fish and roots. The river is three miles wide now.

The fog was thick in the morning and when the fog lifted Clark writes, "Great joy in camp we are in *View* of the *Ocian,* , this great Pacific Octean which we been So long anxious to See. and the roreing or noise made by the waves brakeing on the rockey Shores (as I Suppose) may be heard distictly." **(13)** They were a little ahead of themselves because they were not quite to the ocean yet, but located near Pillar Rock on the north shore of the inlet near Dahila, Washington. The waves grew too high to proceed so they made camp on a narrow area that had barely enough room for them. The wind blew all night and rained. The men were about as wet and cold as you could get. The water was so brackish they used rainwater to drink. They moved their camp from that position up to an area a short distance away; this gave them more protection from the wind and the tide. They shot a few ducks for something to eat. Roasting probably was out of the question as the rain continued. They probably boiled the ducks in salt water. Perhaps a little dried fish was still available.

On November 10, they tried to move and again forced back to shore. The storm just did not to let up. They found an area with a lot of driftwood and built fires to dry their blankets and clothing. Clark said, " we are all wet the rain haveing continued all day, our beding and maney other articles, employ our Selves drying our blankets; nothing to eate but dried fish pounded which we brought from the falls. We made 10 miles today". **(14)** They had described seal, porpoises, and waterfowl but nothing about the hunters shooting anything. The wet conditions were not conducive to firing a flintlock rifle. Their camp was along a steep shore that contained spring water, which they needed. The rain continued and the men had no shelter. They used blankets and the grass mats that they purchased from the Indians as the only shelter, but there probably was not enough for everyone. During the day, a canoe of Indians came by carrying red salmon (sockeye). They purchased a few. The Indians canoed across the bay about five miles in vary large waves. The Indian canoes had higher bows and were made for taking the waves.. It astonished Clark as he would not attempt the waves and thought those Indians could handle a canoe better than anyone he had ever seen. Later that day some of the men found a small creek where they jigged and shot several fresh salmon. It was still raining which makes roasting difficult, so they boiled these fish in fresh spring water for dinner.

The next day hunters hunted in the rain and very strong winds but the banks were so steep they had to return. When their camp became too dangerous, they moved a short distance so they could secure the dugouts. Clark's description tells the peril they were in. "It would be distressing to a feeling person to See our Situation at this time all wet and cold with our bedding &c. also wet, in a Cove Scercely large nough to Contain us, our Baggage in a Small holler about ½ a mile from us, and Canoes at the mercy of the waves & drift wood, we have Scured them as well as it is possible by Sinking and wateing them down with Stones to prevent the emence [waves] dashing them to pices against the rocks— one got loose last night & was left on a rock by the tide Some distance below without recving much damage; fortunately for us our Men are helthy." **(15)**

They again procured salmon from the creek to eat for dinner. These salmon were at the end of the spawn and were of very poor quality. Three men used the canoe they purchased from the

Indians to try to navigate the shoreline above but could not control her. November 13th and the rain continued Whitehouse writes, "Our Buffalo robes are getting rotten, and the most part of our baggage were wet." **(16)**   They again had nothing to eat but pounded salmon as the salmon in the creek are of very poor quality.

It was not until mid afternoon on 15 November, that the rain and wind subsided. The canoes were hastily loaded and they proceeded west along the shoreline and made camp in an area close to Sand Island. They could finally see the entire Pacific Ocean. The men had collected boards from a deserted Indian village they had passed and proceeded to make shelter from the rains. They have made it safely to the Pacific and only Lost Sgt Floyd. It was a remarkable feet for the Corps of Discovery. Now they have to survive the winter on the Pacific Ocean shoreline.

Chapter 11
Winter on the Pacific Coast

It was November 16, 1805. The morning was clear and beautiful. Lewis was ahead looking at the ocean side for ships that were to be in the bay and for a place for a winter quarters. The men could see miles into the ocean and nothing was stirring but high waves. Gass writes, "We are now at the end of our voyage, which has been completely accomplished according to the intention of the expedition, the object of which was to discover a passage by the way of the Missouri and Columbia rivers to the Pacific ocean; notwithstanding the difficulties, privations and dangers, which we had to encounter, endure and surmount." **(1)**

They dried their gear and the hunters were able to secure two deer and several birds for the meals of the day. This is perhaps the first red meat they have had for two weeks. Probably being out of grease and salt, they boiled the meat in water. Not much taste but protein was important, as most of their dried salmon had spoiled from getting wet. The next day Captain Lewis returned to camp with news that he had proceeded to a Point where the visiting ships docked, but they had sailed away. Cape Disappointment is the name given the point overlooking the ocean.

Clark moved down the coastline in the direction from where they came, viewing the area and scouting for a winter camp. When they passed by this area it was raining with low clouds and fog. They were unable to get a look at the land at the time. Deer seemed easy to shoot for the hunters and on November 19, Clark writes of obtaining breakfast on the go. He writes, "after takeing a Sumptious brackfast of venison which was rosted on Stiks exposed to the fire." **(2)** When on the move this was probably the most used method so that each man cooked his own meat by roasting it over an open fire, cooked just the way each man liked it. This method is simpler than roasting with sticks as one does not need to attend to the roasting process as carefully and is less time consuming. Using sticks for roasting greatly increases the chance of loosing the meat in the fire. They were again with the Chinook

Indians and traded for food and clothing. The rain continued and held up the trek west along the shore. The store of trinkets they used for trade was diminishing and the Captains put restrictions on trade.

Where would they put the winter camp? A decision was needed. The Captains called a meeting of the men for their vote, which included Sacagawea and York, but strangely not Charbonneau. Whitehouse writes, "The greater part of our Men were of opinion; that it would be best, to cross the River, & if we should find game plenty, that it would be of an advantage to us, for to stay near the Sea shore, on account of making Salt, which we are nearly out of at this time, & the want of it in preserving our Provisions for the Winter, would be an object well worth our attention." **(3)** Six of the party wanted to go inland back toward the Falls, 10 of the party wanted to stay on the North Side of the River and 12 wanted to cross. The Captains took the party's vote and decided to cross the bay and search for an area along the point.

On November 26, they crossed to the south shore of the Columbia and camped around the area of Svenson, Oregon. The rain again continued and the wind blew. The Clatsop natives wanted to trade for food, but their prices were too high and the reserve of trading material was low. They moved to the John Day River and camped as the wind and rain made it impossible to proceed. There was not any game and they had to subsist on partly spoiled pounded salmon. Fresh water was not available so they had to collect rainwater that pooled in the open canoes. Clark writes on November 28th, "!O how Tremendious is the day" "we are all wet bedding and Stores, haveing nothing to keep our Selves or Stores dry, our Lodge nearly worn out, and the pieces of Sales & tents So full of holes & rotten that they will not keep any thing dry..." **(4)**

Capt Clark and four men took a canoe and continued up the coast to the west looking for a winter quarters. They were able to find game and the landscape was a little less threatening than at the present camp. The availability of food was a big factor in a camp selection as Clark writes, "our diat at this time and for

Severall days past is the dried pounded fish we purchased at the falls boiled in a little Salt water." **(5)**

It was now necessary, although they may not have realized it, to boil their food. They have stated in journal entries in the past few days that some to the fish was spoiled. This means that bacteria was present and boiling prevented what may have been another sickness like they had at the Clearwater. Unfortunately, the next day Clark and a few of the men were sick with diarrhea and weakness. They did not describe the sickness but they probably had eaten some of the dry salmon before they boiled it.

Lewis and Clark were in separate locations and both were looking for food and a place to construct a winter quarters. By December 5, each party had killed several Elk and a few deer. One of the hunting parties returned and said they had found a good place along a river running into a bay that would give sufficient protection from the winds of the ocean and in an area of plenty of game. When the hunters returned to Clark's camp they brought some of the meat with them and some marrowbones that Sacagawea made into a soup using roots and the meat. This was very agreeable to Clark because he had not been able to eat while he had been sick. He wrote, "The Squar Broke the two Shank bones of the Elk after the marrow was taken out, boiled them & extracted a Pint of Greese or tallow from them— " **(6)** Grease must have been at a premium when rendering bones for the grease was a described activity.

Lewis again joined Clark and they were ready to move to the location the men said would be a good place to build quarters for the winter. However, the rain and the wind prevented them from moving until December 7. They paddled against the tide to a deep bay, at the head of the bay was a river which they proceeded to enter; they called it the Lewis and Clark River. They went about three miles then moved their baggage from the canoes to a higher spot that would not be affected by the tide and prepared to call this location winter quarters.

The next day some of the men retrieved several of the Elk that they had killed a few days earlier. Clark and five men set out to find a path to the sea in hopes of making salt for preserving the meat. They found that the Elk spoiled very rapidly and that salt was going to be necessary to preserve meat for winter. Clark joined two Indians who took him to their camp at present day Seaside. They invited him to eat which was a good way to see what the diet of the area might produce. Clark writes, "was treated great Politeness, we had new mats to Set on, and himself and wife produced for us to eate, fish, Lickorish, & black roots, on neet Small mats, and Cramberries & Sackacomey berris, in bowls made of horn, Supe made of a kind of bread made of berries common to this Countrey which they gave me in a neet wooden trencher, with a Cockle Sheel to eate it with. " **(7)**

Clark made his way back to the encampment and found Lewis had the men leveling ground and felling trees in preparation for construction of the fort. It had been raining every day and the weather was damp and chilly. The men were not as strong as they had been in the past as their diets were not nutritious. They had colds and several already had injuries from building the fort.

Shannon and Drouillard being very successful hunting killed 18 elk. They butchered the animals and left them to pickup the next day. The cabins were beginning to take shape. Split logs covered the roofs. They hurried to build a room where the meat would be stored. The meat seemed to be spoiling in about four to five days so it had to be stored and prepared carefully. Keeping it dry would prevent the immediate buildup of bacteria and perhaps keep it a few more days. When they procured enough salt for the meat, it would reduce the process of spoiling a while longer. They were so busy building the fort that they failed to describe the process of drying the meat. The wet weather continued and so did the construction.

It was Dec 16, Clark and the men returned after retrieving meat. The journey was a terrible experience because of the rainy, windy and just plain miserable weather. It was so wet some of the party could not get a fire going and stayed out all night under elk skins. They were sore and many had injuries from the experience. The meat they hung in the new storeroom was now four days old. Clark described the experience, "The winds violent. Trees falling in every derection, whorl winds, with gusts of rain Hail & Thunder, this kind of weather lasted all day, Certainly one of the worst days that ever was." **(8)**

The primary concern to the Captains was preserving the meat before it spoiled. They ordered the meat "fleeced" which is separating the muscles within the meat to smaller pieces. They lit a small smoking fire in the hut to smoke as much of the meat as possible. Keeping it dry was most important and the smoke would prevent spoilage for a short time. The separation of the muscles maintains the muscle membranes around the hunk of meat leaving a barrier that will lengthen its freshness. Cutting the meat allows the area of the slice as a point for the bacteria to begin spoiling the meat. They undoubtedly had to cut many pieces of meat but as much as possible they would leave each separated muscle whole. On Dec 22, Clark confirms the smoking process was not done properly. He writes, "We discover that part of our last Supply of meat is Spoiling from the womph [warmth] of the weather not withstanding a constant Smoke kept under it day and night." **(9)**

The small fires that Clark said were going day and night were not hot enough to kill bacteria. We know today that if the temperature of the inside of the meat did not reach at least one hundred-forty degrees Fahrenheit, the meat would retain moisture and would allow the growth of bacteria. If the fire temperature to bring the meat to one hundred-forty degrees was less, then it might even enhance the ability of bacteria to grow. The small fires that Clark talks about needed to bring the smoke house between one hundred-twelve and one hundred forty degrees for ten to forty minutes. If the temperatures get over one hundred forty degrees the outside surface becomes dry and will not allow the penetration of the smoke and will slow down the curing process. Obviously, the temperature never reached the proper temperature for a cure. **(10)**

The Captains moved into their hut on December 23. The weather remained wet and windy. Occasionally Indians dropped by to trade, mostly roots as the men felt the salmon was spoiled. On Christmas day, they traded gifts from the Captains to the men giving Tobacco to those who used it and silk handkerchiefs to those who did not. It was not a celebrating day as Clark writes, "we would have Spent this day the nativity of Christ in feasting, had we any thing either to raise our Sperits or even gratify our appetites, our Diner concisted of pore Elk, So much Spoiled that we eate it thro' mear necessity, Some Spoiled pounded fish and a fiew roots." **(11)** Ordway and Whitehouse also wrote of the poor food conditions but they both added that the taste of the meat was even less edible in the absence of salt.

The huts were completed, but only the Captains' hut had a chimney. The smoke in their hut from the fires was unbearable. Some of the men proceeded to build chimneys in their huts. The fleas were particularly troublesome so the men thoroughly dried their blankets by the fires to kill or at least drive the pests out of their bedding so they could sleep at night.

On Dec 27, the weather warmed and J. Fields, Willard, Gibson, Wiser and Bratton were sent with five of the largest kettles to make salt at Point Adams. The hunters R. Fields, Shannon, Drouillard, Labich, and Collins were hunting for fresh meat. Clark blames the spoiled meat on the warm, wet weather and them not taking proper care of the meat. This analysis is probably correct to some degree. If the smokers would have been up to temperature, they might have been successful in stopping the bacteria from multiplying in those conditions even though they may have had a head start with the late arrival. Indians arrived at camp, traded dried blackberries, and roasted roots that tasted like licorice. The Captains were delighted with the trade of a few fishhooks and wool in exchange. Because of their deteriorating food supply, they welcomed the taste of something different especially if it was not spoiled.

The Indians told them of a beached whale close to their village. The men in camp continued building the fence around the fort. Indians were coming regularly to trade roots for trinkets. They appreciated the roots since the hunters were not successful finding much meat. The weather continued to be windy, damp to wet and was not conducive to allowing Lewis to advance to the whale for several days.

On Dec 30, the fort was finished. Killing four elk the hunters returned to camp with the meat immediately and hung in the smoke house. In late afternoon, they ate roasted meat, boiled tongue and marrowbones. Clark writes, "we had a Sumptious Supper of Elk Tongues & marrow bones which was truly gratifying." **(12)** The marrowbones laid next to the fire and heated thoroughly, cracked and the warm marrow eaten. The marrow adds the proteins, vitamins, complex b, calcium, magnesium, zinc that they have been missing in their diet. Bone marrow also strengthens the immune system in the body. **(13)** The weather for the month was very wet. Clark's weather journal showed it rained to some extent every day.

On New Years Day, the men fired a volley in celebration. The fort was completed and the camp seemed to have enough meat. They traded for enough roots to keep the camp stocked with something edible other than meat. They were dreaming of next year when they could celebrate a new year in earnest with the rest of the civilized world.

The elk that they were getting are very lean and evidently do not taste very good. It seems they were not very satisfying. Clark wrote in this journal on January 3, 1806. "our party from necescity have been obliged to Subsist Some length of time on dogs have now become extreamly fond of their flesh; it is worthey of remark that while we lived principally on the flesh of this animal we wer much more helthy Strong and more fleshey then we have been Sence we left the Buffalow Country. As for my own part I have not become reconsiled to the taste of this animal as yet." **(14)** The Indians brought three dogs to trade which probably inspired Clark to write this statement.

Willard and Wiser have not returned from the ocean with any salt; it has been six days since they left. The Captains feared for their fate so Lewis sent Gass and Shannon to look for them. Late that afternoon, Willard and Wiser arrived at camp not having crossed paths with Gass and Shannon. They brought with them samples of the salt they made, which was about three quarts to one gallon each day. They also brought a good portion of blubber from the whale. Lewis describes, "it was white & not unlike the fat of Poark, tho' the texture was more spongey and somewhat coarser. I had a part of it cooked and found it very pallitable and tender, it resembled the beaver or the dog in flavour; it may appear somewhat extraordinary tho' it is a fact that the flesh of the beaver and dog possess a very great affinity in point of flavour. **(15)**

Lewis seems as though he would prefer anything that tasted fatty but the blubber reminds him more of the dog and beaver. They are not his most favorite of foods with or without salt, but they are better than lean elk or deer. The men tasted the salt and were very pleased. Clark decided to take two canoes and 12 men and Sacagawea, and go to the whale and retrieve some of the blubber. It was a two-day trip as the difficult terrain and bad the weather prevented a good passage. They passed the salt makers and proceeded along the coast over mountains to Ecola Creek in the area of current day Cannon Beach. The next morning Clark and his crew visited the whale to find the Indians were busily rendering the blubber and making oil. Clark had hoped to get a supply of blubber without having to buy it, but there was not much left of the whale but a skeleton. Clark tried to trade for some blubber but he did not bring enough trading material and the Indians would not trade reasonably for the trinkets he brought. He was only able to acquire about 300 pounds of blubber and a few gallons of oil. Clark was happy to secure what he did since they had been out of oil or grease for some time. Clark said he was very weak for lack of proper food and figured it was going to be a slow walk back over the mountain to get to the canoes and the salt makers. They dined on fresh elk that evening and proceeded to the fort the next morning.

At the fort Lewis was low on food and wrote, "some marrowbones and a little fresh meat would be exceptable; I have been living for two days past on poor dryed Elk, or *jurk* as the hunters term it." **(16)** His hunters were successful the next day securing elk.

Hunters searched daily for elk. It was difficult hunting, but they did find a herd and they usually shot more than one animal. They may have gone for several days without finding anything, then locate a herd and kill several animals.

On January 12, they shot seven. The Captains found that the sporadic acquisition of the meat made it necessary to make a system to ration the meat to the messes. Clark writes, "We have heretofore devided the meat when first killed among the four messes, into which we have divided our party, leaveing to each the Care of preserving and distribution of useing it; but we find that they make such prodigal use of it when they happen to have a tolerable Stock on hand, that we are determined to adapt a Different System with our present stock of Seven elk; this is to jurk it and issue it to them in Small quantities." **(17)** Six men worked cutting the meat into strips and hanging it on a scaffold. They built the fires and began to jerk the meat so that it would not spoil.

Lewis was now catching up on his journals. He described the many tribes along the river, their habits, clothing, and ways to make a living. The reason the men had been able to secure elk in the area is that the Indians had an inferior way to hunt elk. Lewis described killing elk with pieces of the arrow still in the flesh indicating that they were hunting the elk but had a poor kill ratio. The Indians had traded for muskets but not being rifled, as most of the Corps guns, they were smooth bore muskets and not very accurate. In fact, as they ran out of lead, they used rocks and small pebbles for shot. They were unaware of the damage to a musket barrel from using stones, thus making the gun even more inaccurate.

On January 16, I believe, the Corps had become more relaxed knowing that game was available and easily secured by a good hunter. The natives were close by to trade, and the methods used to preserve meat prevented spoilage and they had dry shelter. The men are making better clothing from the hides they have retrieved from hunting, trapping or trading. Lewis writes, "we have plenty of Elk beef for the present and a little salt, our houses dry and comfortable, and having made up our minds to remain until the 1st of April, every one appears content with his situation and his fare." **(18)**

Lewis continues to write about life among the Indians and writes about the cooking articles used by the natives. He writes, "The Culinary articles of the Indians in our neighbourhood consist of wooden bowls or throughs, baskets, wooden spoons and woden scures or spits. Their wooden bowls and troughs are of different forms and sizes, and most generally dug out of a solid piece; they are ither round or simi globular, in the form of a canoe, cubic, and cubic at top terminating in a globe at bottom; these are extreemly well executed and many of them neatly carved the larger vessels with hand-holes to them; in these vessels they boil their fish or flesh by means of hot stones which they immerce in the water with the article to be boiled. They also render the oil of fish or other anamals in the same manner. Their baskets are formed of cedar bark and beargrass so closely interwoven with the fingers that they are watertight without the aid of gum or rosin; some of these are highly ornamented with strans of beargrass which they dye of several colours and interweave in a great variety of figures; this serves them the double perpose of holding their water or wearing on their heads; and are of different capacites from that of the smallest cup to five or six gallons; they are generally of a conic form or reather the segment of a cone of which the smaller end forms the base or bottom of the basket; these they make very expediciously and dispose off for a mear trifle; it is for the construction of these baskets that the beargrass becomes an article of traffic among the natives; this grass grows only on their high mountains near the snowey region; the blade is about $\frac{3}{8}$ of an inch wide and 2 feet long smoth pliant and strong; the young blades which are white from not being exposed to the sun or air, are those most commonly employed, particularly in their neatest work. Their spoons are not remarkable nor abundant, they are generally large and the bole brawd; their meat is roasted with a sharp scure, one end of which is incerted in the meat with the other is set erect in the ground; the spit for roasting fish has it's upper extremity split, and between it's limbs the center of the fish is inscerted with it's head downwards and the tale and extremities of the scure secured with a string, the sides of the fish, which was in the first instance split on the back, are expanded by means of small splinters of wood which extend crosswise the fish; a small mat of rushes or flags is the usual plate or dish on which their fish, flesh, roots or burries are served. they make a number of bags and baskets not watertight of cedar bark, silk-grass, rushes, flags and common coarse sedge; in these they secure their dryed fish, roots, buries, &c." **(19)**

On January 20, Lewis reported that their jerked meat was running out and he dispersed the hunters. In a short time, Shannon and Labich shot two elk. However, when they got the elk to camp they were of very poor quality, thin and lean. Their salt was gone. They used it a few days back, probably by soaking the raw meat in salt brine to break the meat cells and allow it to jerk faster and stay edible for longer.

Their tanning process for the hides required brains from the animals. They had not secured enough to complete all of the hides. The hunters retrieved a few more brains with their next kill.

The weather continued to be wet with a little rain almost everyday. On January 24, Lewis reported that it had snowed and built up on the ground. The snow is light but kind of a hail as some of the men report. It was the first freeze they had during the winter. The men at the salt kiln were having a hard time finding game and had to subsist on whale blubber obtained from the Indians. By January 26, five inches of snow had fallen on the level. The cold freezing temperatures continued. Two men traveled to the salt makers with more trinkets for them to use to trade for food. They had been able to make about 3 bushels of salt by this time.

The salt kiln was a simple design. It was a long mound of rocks with the kettles set down in the top with a horseshoe shaped opening to build fire underneath. This burned very hot, heated the rocks and boiled the water. The water evaporated and the salt remained.

The hunters from the fort managed to kill ten elk, but the snow and hard terrain allowed them to retrieve only two of them. Retrieving two was not a pleasant task as the cold wet weather persisted. Whitehouse complained of frost bitten feet. They again have a little meat in camp. Lewis pens, "our fare is the flesh of lean elk boiled with pure water, and a little salt; the whale blubber which we have used very sparingly is now exhausted. On this food I do not feel strong, but enjoy the most perfect health; a keen appetite supplys in a great degree the want of more luxurious sauses or dishes, and still render my ordinary meals not uninteresting to me, for I find myself sometimes enquiring of the cook whether dinner or breakfast is ready." **(20)**

It is now February 1 and Lewis continues to write about the life among the Indians and describes the art of making canoes. Some of the canoes they made were as long as 50 feet and could carry a very large load. All were light to carry, well constructed with a chisel and decorated with figures on the bow. Lewis also took an inventory of the only means of securing a safe trip home, powder and lead. He designed a method of taking the powder in lead canisters. Each canister contains four pounds of powder and 8 pounds of lead carefully sealed to keep the powder dry. They had twenty-seven of the best powder and seven of musket powder. If it were not for the canister design, they would have not had any dry powder at this time. This was truly a life saving invention.

The hunters were hunting daily for fort and the salt makers. On February 5, they killed several elk and proceeded to take them to the fort. The tide had taken away one of their canoes about 10 days before and they found it in a ravine. The weather was moderating and there seemed to be less trading among the Indians. On returning to the downed elk the next few days, they found that some Indians had stolen seven of the elk. Not being able to prevent this kind of thievery, they acted as though things like that happened and did not pursue as it may disrupt the peace. They continued to hunt and upon every successful hunt, they dined on marrowbones and brisket; always looking for the fattest meat, they could find. They devoured the best cuts first, and then jerked the rest of the meat it would not spoil. Hunters were out daily and on occasion would retrieve a beaver. The beaver was a bit fattier and continued to be their meal of choice. On February 12, a Clatsop Indian arrived at the fort with three dogs. With the dogs, he confessed that they had stolen the meat a few days earlier. The apology of sort was fine with the Captains and peace prevailed. The dogs ran off and the gesture was all that remained.

The men working at the salt camp became sick and bored. Bratton and Gibson had lost a lot of weight and had severe winter colds. They returned to the fort where they would be warm and dry to recuperate. The Captains estimated that three bushels of salt was all that was necessary for them to make the trip back to St Louis. Gass writes, "One of the men brought word from the salt works, that they had made about four bushels of salt; and the Commanding Officers thought that would be sufficient to serve the party, until we should arrive at the Missouri where there is some deposited." **(21)** Ordway and his men from the salt works arrived a little past midnight at the fort with the kettles and salt. They were becoming sick and very weak. Gibson, Willard, Ordway and

Bratton may have had a virus. Lewis wrote on February 22, "the general complaint seams to be bad colds and fevers, something I beleive of the influenza. " **(22)**

Shannon and Labich thought the elk had left the area. They were not finding the herds they found previously and it was almost impossible to shoot enough for camp. The Indians came by and traded sturgeon and small fish they caught in nets farther up the Columbia. The sick were healing and never had the men been as well dressed in leather hides as they were now. The weather is still wet, warming slightly. It is February 26 and their stock of provisions is getting low. They sent Drouillard, Cruzatte and Weiser up the Columbia to trade for sturgeon, small fish and to purchase breadroot. The hunters went to different areas in search of game. They found elk but at a distance of about ten miles from camp; they were successful in returning with several.

On March 2, Drouillard, Cruzatte and Weiser returned to the fort with a good supply of sturgeon, fresh small fish and wapato roots they had purchased. The men welcomed the change from their diet of lean elk meat. In order to keep the small fish from spoiling, they used the Indian method by hanging the fish by the gills on small sticks and dried in the smokehouse. The sturgeon was steamed. Lewis describes the technique, "the fresh sturgeon they keep for many days by immersing it in water; they cook their sturgeon by means of vapor or steam. The process is as follows. a brisk fire is kindled on which a parcel of stones are lad. when the fire birns down and the stones are sufficiently heated, the stones are so arranged as to form a tolerable level surface, the sturgeon which had been previously cut into large fletches is now laid on the hot stones; a parsel of small boughs of bushes is next laid on and a second course of the sturgeon thus repating alternate layers of sturgeon and boughs untill the whole is put on which they design to cook. it is next covered closely with matts and water is poared in such manner as to run in among the hot stones and the vapor arrising being confined by the mats, cooks the fish. the whole process is performed in an hour, and the sturgeon thus cooked is much better than either boiled or roasted." **(23)**

By March 6, finding game was almost impossible. Lewis had the canoes pulled and repaired so that if the hunters did not find game in the next day or two they might have to leave Clatsop and go along the Columbia to find meat. He sent a couple of men upstream to buy more sturgeon, fish and roots but this would not sustain them until April 1st when they planned to leave. They were lucky in finding two elk and returned with them to camp. The men returned from upstream with sturgeon, and fresh fish and roots. Many of the men were out hunting daily and finding less elk sign every day. The men were becoming restless. Some of them were still sick and Bratton was almost bedridden with a bad back. They would soon have to decide whether to begin their return trip.

Looking over their return trip stock Clark writes, "Our party are now furnished with 358 par of Mockersons exclusive of a good portion of Dressed leather, they are also previded with Shirts Overalls Capoes of dressed Elk Skins for the homeward journey." **(24)**

The camp was out of meat again, but the hunters brought only a couple of elk. Captain Lewis sent Drouillard to the Clatsop village in hopes of buying a canoe. They had lost a couple of their dugouts when the wind broke the leather tie ropes and swept them away. Drouillard returned not able to purchase a canoe. Nothing was more valuable to the Indians than their canoes. They did trade for hats and roots. The hunters continued to be marginally successful in securing almost enough to eat day by day.

On Mach 16, Drouillard went to the Cathlamah Villiage to see if they would trade for a canoe. When Drouillard returned, they found that he had found the canoe lost a few days ago and had traded one of the Captain Lewis' coats and a half a carrot of tobacco for a good canoe. The natives would take nothing less than the coat and since they were in dire need of another canoe, they traded. They needed another canoe and thought that they might have to take one from the Clatsop Indians who had stolen their elk meat previously. The next day, Ordway wrote in his journal. "4 men went over to the prarie near the coast to take a canoe which belongd to the Clotsop Indians, as we are in want of it." **(25)**

They were preparing to leave, and left the Indians a list of the men on the expedition and told them to show this to any other civilized people that may enter the area. The canoes needed sealing in some spots to be water worthy but the rains continued and they could not complete the task. An elk delivered to camp supplied them with food for another evening. The next day it rained hard again. They had little provisions remaining and hoped that the weather would break and allow them to finish repairing the boats so they could depart.

On March 23, 1806, the Corps of Discovery departed Fort Clatsop and headed back up the Columbia toward St. Louis. Clatsop was good to them once the construction was complete. They were able to kill around 150 elk and about 20 deer. Trading with the natives was successful and enough salt collected for the return trip. A good supply of clothing was prepared. Hunters went out ahead of the main party to procure some game for later in the day, as they had nothing to eat. Their worst problem was the almost daily rain and continuous cold winds. The one thing they would not miss was the state of constant wetness. Their only failure was not finding ships of a civilized culture. These ships would surely have had stores of rations the Corps could have obtained and possibly made their return trip by sea rather than retracing their land route.

Fort Clatsop served as the Corps of Discovery's quarters during the winter of 1805-06. Before departing, the Captains posted a roster of the men of the Corps on the wall of the fort. They gave the fort to the Clatsop Indians. This reconstructed fort shown here burned to the ground as the Lewis and Clark bicentennial (2003-2006) was drawing near. A new Fort was built immediately after some finger pointing and sensationalism on the cause of the fire.

Chapter 12
Fort Clatsop to Travelers Rest

On the afternoon of March 24, the Corps of Discovery was on its way back. The Captains minds focused on the journey ahead and the perils that lie before them. The one thing that greatly concerned them was the availability of game. They knew the Indians would be along the shores of the Columbia, but Indians would have the usual poor quality fish and roots. The spring spawn of fish moving up the river would soon begin then there would be fresh fish to eat. They keenly remembered the shortage of firewood and the absence of game. Both Captains suspected the Indians probably had harvested nearly all the game from the area. Lewis laments the shortage of migratory birds, which have not started to move north. A keen eye for game and dependence on the hunters was important. The weather played an important role during the trip to the ocean. It would be important on the trip back. The Captains now considered the ever-changing terrain that sometimes was so thick with timber it was impossible to walk through, then the barren plains or the hills and mountains so steep that the hunters could not climb to hunt. All these factors were important factors in selecting camp locations.

They stopped at one of the first Indian encampments across from Karlson Island. Lewis purchased a dog and some roots as backup food for Bratton and Willard who were still recovering from injuries and sickness. They moved on and camped northeast of current day Brownsmead at Aldrich Point where they had a quick breakfast of a little leftover elk meat. They proceeded on and tried to buy some fish from a group of Indians, but the trade failed because the prices were far too high. As the Expedition moved up the river, the going was slow due to the wind and a very strong river current. They found a place to camp along a creek north of present day Clatskanie, Oregon. Close by were some Indians with fresh sturgeon and seal. They gave the Captains some loin meat of the seal. Clark writes, "they gave us Some of the flesh of the Seal which I found a great improvement to the poor Elk." **(1)**

The next day they purchased a sturgeon, as the hunters were not finding much game other than a few birds.

On March 29, they continued to move up the river and found many Indians wishing to trade. Some of the trades included dogs, roots and Sturgeon. The hunters were successful on Deer Island and killed several deer. There was fresh meat again for camp and there was little complaining of hunger among the men. Willard is now well and Bratton feeling better by the day. They continued up the river only making a few miles each day. They came to an area that the natives said was abundant with wild life. A beautiful valley that Lewis describes as follows, "this valley is terminated on its lower side by the mountanous country which borders the coast, and above by the rainge of mountains which pass the Columbia between the great falls and rapids of the Columbia river. it is about 70 miles wide on a direct line and it's length I beleive to be very extensive tho' how far I cannot determine. this valley would be copetent to the mantainance of 40 or 50 thousand souls if properly cultivated and is indeed the only desireable situation for a settlement which I have seen on the West side of the Rocky mountains." **(2)** This valley now comprises the cities of Vancouver, Washington and Portland Oregon.

The Expedition moved passed large encampments of Indians to an area by the Sandy River. Several Indians stopped by the encampment and complained that the Salmon had not yet arrived at the falls above and their people were in want of food. Their winter supply of salmon was gone. They had dogs but Lewis could only visualize the condition of the dogs as poor if there was not enough food for them. Lewis writes on April 1. "This information gave us much uneasiness with rispect to our future means of subsistence. Above falls or through plains from thence to the Chopunnish there are no deer Antelope nor Elk on which we can depend for subsistence; their horses are very poor most probably at this season, and if they have no fish their dogs must be in the same situation. under these circumstances there seems to be but a gloomy prospect for subsistence on any terms; we therefore took it into serious consideration what measures we were to pursue on this occasion; it was at once deemed inexpedient to wait the arrival of the salmon as that would detain us so large a portion of the season that it is probable we should not reach the United States before the ice would close the Missouri; or at all events would hazard our horses which we lelft in charge of the Chopunnish_ who informed us that they intended passing the rocky mountains to the Missouri as early as the season would permit them wich is as we believe about the begining of May." **(3)**

This area looked promising for hunting, with an abundance of game. Hunters changed direction to evaluate the area and secure some meat for the camp. Late in the afternoon, they returned with four elk and two deer. The hunters said there was abundant game and this would be a good place to put in a store of meat. Captain Lewis agreed to stay in this area to lay in a supply of dried meat for their trip. The Captains assigned work groups for hunting, hauling and wood gathering and building scaffolds to dry the meat, or to cut the meat in strips for drying. The Indians kept coming to the camp; sometimes three or four canoes at a time. They became so troublesome that Lewis sent some of the men elsewhere to dry the meat. The starved Indians picked up pieces of discarded meat and bones for their own use.

Clark and a few men explored the Willamette River and discovered that some of the Indians along the way had endured an epidemic of small pox that nearly wiped the tribes out. They killed several elk and deer on their excursion and dried the meat. When the party returned to camp Lewis ordered it cut into thinner strips as he felt that it was not dry enough and it would quickly spoil. The wet weather did not help dry the meat. They probably had to build fires large enough to create a temperature high enough to partially cook the meat to keep it from spoiling. They continued to hear from the natives that the tribes by the falls were starving for lack of food, which made this collection of meat very serious. Since there was little game ahead, drying the meat thoroughly was necessary so that it would not spoil.

On April 6, they moved their camp upstream a few miles. The hunters had killed several elk, which they placed on scaffolds to dry. Because of the damp weather, fires continued burning under the scaffolds and tended all night. Clark wrote, "we derected that fires be kept under the meat all night." **(4)** The wind blew so violently the next day that they remained in camp. Lewis had the meat re-dried as a precaution against spoilage.

The Expedition packed all their dried meat and a few dogs they had purchased from the natives and continued upstream. They camped the night in the area below the now Bonneville Dam at Bradford Island. Along the way, they mentioned the beautiful cascade of Multnomah Falls.

The next day they pulled most of the canoes out of the river and carried some of the baggage to the top of the falls. The main problem was that a few of the Indians were thieves and the Expedition could not trust them. A couple of renegades stole a tomahawk and made several attempts at other articles. The men were prepared to kill any of them if thefts happened again. The chief was apologetic and said that it was a couple of their undisciplined men. They continued the portage the next morning, but lost one of the canoes, leaving them with only four. It rained all the next day causing the Expedition to spend three days getting over the cascades.

Lewis visited a tribe of Indians and did some trading. He traded some of the hides they tanned during the winter for two small canoes and a few dogs. Lewis was beginning to like dog, he writes. "the dog now constitutes a considerable part of our subsistence and with most of the party has become a favorite food; certain I am that it is a healthy strong diet, and from habit it has become by no means disagreeable to me, I prefer it to lean venison or Elk, and is very far superior to the horse in any state." **(5)**

They continued to move upstream, killing a deer or two daily, which was just enough food to keep them from being terribly hungry. They also were bartering with the natives buying an occasional dog. As they approached the area of Mill Creek, now The Dalles, they noticed many horses around the encampments of the Indians. They tried to buy some horses to assist in the trip from several different tribes no horses could be secured. After Lewis told Clark to double the offer and he was able to secure four horses. The Indians that Clark was visiting were not finding any salmon moving and were eating their dried salmon from last season. Clark purchased a couple of dogs and proceeded up the river to see if he could trade with another tribe. There was no firewood and almost no game in the area. Lewis who was on the south shore across the river from Clark had his men make twelve packsaddles for the horses.

On April 18th Lewis crossed the river to Clark at the Indian village and found that Clark had purchased four horses. They proceeded to move the baggage and portaged one canoe. Clark was above the falls and shuttled the horses to the bottom along with the dogs he purchased. They broke up two canoes for firewood and cooked the dogs. Lewis finally made a deal by trading one of the cooking kettles for a horse. They used the four horses to transport the baggage to the top of the falls. They really had not planned to move by boat any longer, but trading for horses was not working out. During the evening, the Indians caught the first salmon of the season. The Indians of the area had a tradition that required issuing a small piece of the salmon to the children to make the fish run continue and be strong.

Clark finally traded two kettles for four more horses. This was a serious trade. Lewis writes, "we wer obliged to dispence with two of our kettles in order to acquire those; we have now only one small kettle to a mess of 8 men." **(6)** These kettles are the lifeblood of the mess. Without a kettle, they would not be able to cook with grease or boil fish in water or even secure water from the stream. I believe the kettle, next to the gun, was one of the most valuable pieces of survival equipment on the journey. Other items that may hold water were probably available but not like a brass kettle, that holds the heat for cooking.

Since Lewis did not trust the natives, he had guards posted by the horses to make sure they would not run away or be stolen back by the Indians. Willard was not paying attention and one of the horses got loose and ran away. Lewis was outraged, and reprimanded Willard more than he had ever done before. As it worked out the horse returned the next day.

Lewis was at his wits end with this tribe of Indians. They were thieves that would steal anything. Lewis lost his temper and threatened to kill the next Indian that stole anything. They now had ten horses and continued up the river to meet Clark who was bartering for more horses. Before they left, they again purchased a couple of dogs and roasted them. They used the remains of the broken canoe for firewood.

On reaching the next camp, they found Clark was unable to secure more horses and had not eaten much since he left. The Indians had nothing but dried salmon and old roots and seeds. They were rationing their food and only small amounts were available. The Expedition continued but had constant trouble with their horses. They purchased a few more horses. It was spring and ten of the thirteen horses they had were studs and were easily riled. The only food available was dog. However, no matter what the price, that was their only food except roots of different types and a few acorn nuts. Unfortunately, there was no wood and the Indians wanted more for the wood than the dogs.

On April 24, they sold their canoes to the Indians and purchased more horses. They were now prepared to travel totally on land. Since wood was not available for cooking fires, the Expedition was forced to use grass and dried willow branches for fire to roast their dog meat, ducks, snakes and whatever small game might be available. They were back in the country of the Walla Walla, Umatillas and Nez Pierce Indians, near Alderdale, Washington. Their food supply was short so they resorted to boiling and eating some of the dried elk meat. On April 28, they met some of the Indians they had met on the way to the ocean the year before. The chief traded a beautiful white horse to Clark for his sword and a few balls and powder along with some other articles. The Chief was pleased and so was Clark.

They were on the north side of the Columbia and the Indians assisted in getting the horses to the south side of the river. The river in this area must have been shallower and allowed the horses to swim. They waited until the next morning to move their baggage across the river. That evening there was a great celebration with dancing and fiddle playing in camp. Large numbers of Indians also danced for them. Realizing they must procure some staples to get over the mountains, they bought several dogs and a quantity of roots.

The next day they proceeded on the south shore of the Columbia up Walla Walla River toward the mountains. This was the first night they had the luxury of plenty of firewood. The

Indians that were with them would not eat the dog that they had prepared. They preferred the meat of an otter the hunters had killed. They now had 23 horses to make the trip across the mountains. They proceeded north in a direction that bordered a mountain range, staying at its base to avoid traveling through the mountains to get to the Snake River. The Corps followed an Indian trail that would lead them to their destination over a terrain that was much less treacherous and promised to be more bountiful with game than the mountain trail. However, they were only able to kill a deer. On May 3rd Lewis wrote, "we divided the last of our dried meat at dinner when it was consumed as well as the balance of our dogs nearly we made but a scant supper and had not anything for tomorrow." **(7)** Their dried meat lasted about a month.

The thorough drying of the meat that Lewis had done in the area of the Willamette River apparently was successful since they said nothing of poor quality or spoilage. Their Indian guide convinced them that they would be at an Indian camp the following day about noon. At noon the next day they arrived at the same village they had visited on October 11, 1805. The occupants were very poor and the purchase of two dogs was difficult. The salmon had not yet reached this location and their winter supply of food had run low. All they had were cous roots to live on until the salmon arrived. The Corps purchased some of the roots, made soup, and ate the dogs. They were now at Lewiston, Idaho

The next day they ascended the Clearwater River, passing through several Nez Perce encampments. They camped with a group at present day Arrow, Idaho. These Indians had several sick people and Clark traded some of his doctoring techniques for a young horse. The next morning the Chief as promised produced the horse, which they immediately butchered for food. Clark continued doctoring the sick and received in payment a young colt for food as they crossed the mountains. However, after they traveled several miles the colt broke loose and ran away. The men probably were disheartened about the escape because they knew that they probably would need the meat.

They continued up the Clearwater and found out that there was still too much snow to pass over the mountains. They had to wait for it to melt, so they continued a short distance to a small creek with abundant game sign and encamped. The dispatched the hunters the next morning to kill some game. It was May 8th and the hunters were somewhat successful in obtaining a few deer. The Indians who were with the Corps also dined on venison and horsemeat. Lewis wrote, "we gave this cheif and the indians with us some venison, horsebeef, the entrels of the four deer, and four fawns which were taken from two of the does that were killed, they eat none of their food raw, tho' the entrals had but little preperation and the fawns were boiled and consumed hair hide and entrals. these people sometimes eat the flesh of the horse tho' they will in most instances suffer extreem hunger before they will kill their horses for that purpose, this seems reather to proceede from an attatchment to this animal, than a dislike to it's flesh for I observe many of them eat very heartily of the horsebeef which we give them." **(8)** They moved up the river to the lodge of the Indian that was taking care of the horses they left last fall. He explained that the horses were in various places, some were days away.

They were camped in the area of Ahsahka, Idaho. Lewis described the country as a fertile valley that could be a very productive agricultural area. It had an abundance of the plants collected, dried and processed into the staple that kept the Indian tribes fed during the winter. This was the Quawmash or the Blue Camas, *Camassia quamash,* and the Cous Bisquit-Root, L*omatium cous.* Lewis described the plants on May 9 he wrote, "the cows is a knobbed root of an irregularly rounded form not unlike the Gensang in form and consistence. This root they collect, rub of a thin black rhind which covers it and pounding it expose it in cakes to the sun. these cakes are about an inch and ¼ thick and 6 by 18 in width, when dryed they either eat this bread alone without any further preperation, or boil it and make a thick muselage; the latter is most common and much the most agreeable; the flavor of this root is not very unlike the gensang. This root they collect as early as the snows disappear in the spring and continue to collect it untill the quawmash supplys it's place which happens about the latter end of June. the quawmash is also collected for a few weaks after it first makes it's appearance in the spring, but when the scape appears it is no longer fit for use untill the seed are ripe which happens about the time just mentioned, and then the cows declines. The latter is also frequently dryed in the sun and pounded afterwards and then used in making soope." **(9)**

That evening the Indians returned twenty-one of the lost horses to Lewis and most of the cached saddles. Most of the horses were in good condition except a few with sore backs from having had excessive riding. Gass made a note in his journal that there were more horses in this area than anywhere he has seen. They also saw that the horses of the Nez Perces are in good condition and well taken care of. Over night, it snowed eight inches.

They proceeded to the area of Kamiah, Idaho to meet with the Indians they had left an American flag with when they passed last fall. The Chief still displayed the flag at the camp. He was very friendly and provided them with a place to camp and food. They gave Captains a couple of bushels of roots and cakes of cous roots and a dried salmon. Lewis told the chief that his men were very hungry, but the roots would not agree with them. He offered as many young colts as they needed for food, but was not happy with the idea because they did not eat their horses unless it was an emergency. This was perhaps the greatest gesture of hospitality that the Expedition had received during the entire journey. Lewis explained that they would not kill the second horse for food until they were totally out of meat and only if it was necessary. They moved their camp to the north side of the Clearwater to wait for the snow in the mountains to melt. There seemed to be good feed for their horses and the Indians said this area had plenty of game. The Captains called this location Camp Chopunnish.

The hunters were successful and killed several bears in the area. The Corps proceeded to render the fat to obtain as much bear oil as they could get for the return trip. They were able to obtain enough venison and purchase enough roots to keep them going without having to kill a horse to survive. The salmon run would be moving into this area in the next few days and that would help the supply of fresh food. The Indians would visit from time to time and on May 14, Lewis shared some bear meat with them. (See Appendix I for Lewis description of how this meat was prepared)   The slow roasting tenderizes the meat and makes it very delicious, but you can get around the strong flavor of the branches placed to keep the meat from the dirt.

The Captains' shelter was rotten and new shelters constructed to protect their gear from the rain. The leather teepee, hauled all the way from Fort Mandan had finally become unusable. Some of the blankets and tanned hides were also tattered and worn and needed replacing.

Hunting was unpredictable. Some days they were successful, but usually there was a shortage of meat to feed the entire camp. They killed troublesome stud horses in want of meat. They also looked for ways to make the meat taste better, as well as for roots and greens that would make the meals more fulfilling. Clark writes, "Shabonos Squar gatherd a quantity of *fenel* roots which we find very paleatiable and nurishing food. The Onion we also find in abundance and boil it with our meat." **(10)**   This was definite proof that at this point they were probably saving their bears oil of which now they had several gallons. Placing onions in the water with boiling meat will add to the flavor. If they were boiling in oil, the onions would have burnt and would have added no flavor to the meat. Now that they have time on their hands, the use of techniques requiring more time were typical. They were saving the rapid cooking methods for the travel time.

It was Lewis's plan that when the snow subsided they would not move slowly over the mountains but would push hard and get over the terrain as fast as possible. On May 22, their meat supply from hunting was exhausted. Lewis ordered a fine colt killed for meat and a few hours later the hunters come to camp with five deer. They now had three colts reserved for meat for the trip through the mountains so they would no longer kill horses for meat. They hoped the salmon run would bring a supply of food up the river shortly. Sacagawea also laid in a supply of roots but not enough for the entire party. They continued to purchase roots although they were running low on trading material. On May 27, they again ran out of meat. One of the hunters found a horse at large, and returned it to camp where it was butchered, then later the hunters arrived with five deer. There were signs of the salmon beginning to run the river but the number was very small and the catch poor. Several days passed with successful hunting. The snow was now melting rapidly and the rivers were violently high and rapid. They caught and brought to camp a few salmon even though they were not abundant. Lewis took note that they now had sixty-five horses in their possession. This should be ample to make the trip over the divide.

On June 10, they left camp and headed toward the Camas Flats in hope of doing some hunting before they begin to climb out of the valley over the mountain. This is a sight. Every man has his own horse and a packhorse lightly loaded and a few extras, some for food and others as replacements for those that may be injured. They were able to secure extra camas roots from the Indians but never were able to get any salmon for the trip because the run did not materialize.

The men camped in the area of Weippe, Idaho where they met the Nez Perce on September 20, 1805. One of the hunters killed a deer. They were also able to kill several ground squirrels. Lewis wrote, "we find a great number of burrowing squirels about our camp of which we killed several; I eat of them and found them quite as tender and well flavored as our grey squirel." **(11)**

Lewis sent out the hunters the next morning to obtain meat for the trip. As he sat in camp looking at the beautiful Camas field in full bloom, he described in detail the procedure that the Indians use in preserving this prairie staple for use during the winter. (See appendix 1 for his description)

The hunters secured about ten deer and returned to camp. They were ordered to dry it but in a different way than normal. Lewis writes, "we directed the meat to be cut thin and exposed to dry in the sun."**(12)** When they were preparing meat for a trip, they usually dried it toughly with fire. However, this time they just wanted the blood to be dried. The meat would spoil much more rapidly than before, but since they did not have a lot of meat, they probably assumed it would be gone before spoilage actually started. They also had saved the bears grease, which made it much more palatable and quicker to cook.

On June 15, they gave up waiting for the Indian guides that were to assist them in finding the trail to travelers rest. The Expedition loaded up and headed into the mountains. The lower elevations did not seem to be covered with snow and they camped in areas where there was enough grass to feed the horses. However, two days later in to the mountains, the steep terrain was covered with snow, some of it very deep. They lost the trail and returned where there was feed for the horses. They cached some of their equipment and baggage then returned to the first night's camp. From there they sent Drouillard and Shannon down to locate the guides that were late coming to the Camas Flats. It now was going to become a challenge for the hunters to supply enough meat to satisfy the Corps until the two men return with guides.

The next day they were successful and got three deer. That was enough to keep the camp somewhat happy. Lewis said they were out of salt except for a little he planned to take with him to Travelers Rest. They moved closer to the Camas flats while waiting for the return of the two men sent in search of the Indian guides. Two other Indians heading up the trail told the Captains they had seen Drouillard and Shannon, and they had no explanation for their delay. The hunters were successful in obtaining enough deer and fish to keep the men going and there was enough feed for the horses, so they remained a little longer in wait.

In the afternoon of June 23, Shannon and Drouillard showed up with three Indians to help guide them over the mountains. They killed several deer for food and in fact had a little surplus to make the trip. Although it was not dried, it would last for another day. They camped the next day at fish Creek at the location of the last good grass for the horses. As they continued the next day to Hungry Creek, they found the snow was beginning to melt rapidly and the passage ahead was probably going to allow them to proceed. Upon picking up their cached baggage, they continued on to Bald Mountain where there was plenty of good grass for the horses. Some of the venison and a few roots was all the food that remained. The hunters had no success obtaining anything but a few grouse.

The next day they made twenty-eight miles and camped in the area of Packer Meadows. Their food was virtually gone. Lewis writes, "our meat being exhausted we issued a pint of bear's oil to a mess which with their boiled roots made an agreeable dish."

(13) They probably boiled the roots again to form a paste like mashed potatoes and then added the bears grease like you would butter to your mashed potatoes. The bear grease would add flavor and give the roots more nourishment.

The Expedition continued and stopped the next day about 13 miles from Packer Meadows in an area where grass was plentiful. The horses were able to satisfy their hunger; the men however, continued with roots and grease. It was twenty-ninth of June, when they finally reached the Hot Springs at Lolo. The hunters were able to secure a couple of deer and the men amused themselves in the hot springs throughout the evening. The following morning they proceeded on and made it to their destination at Travelers Rest.

The trek across those mountains was one that Lewis said he would not forget. The mountain put a group of men at the edge of survival having little game, absence of grass for the horses, and downed timber with deep snow and cold temperatures, which made the passage almost unbearable. The Captains both commented that it was one of the hardest parts of the entire expedition. The river valleys were a welcomed sight.

Buffalo hide or leather tipis worked well on the prairies where the humidity is low, but the Captains discovered why they were little use in the northwest where high humidity soon rots the leather.

Chapter 13
Travelers Rest and Onward

The Captains decided to stay at Travelers Rest so they could build up a supply of enough food for the next part of the journey. The hunters were successful in killing more than a dozen deer in a couple of days, most of which they sun-dried so that a supply could be taken with them. They were not as particular with thorough drying as they had been on the Columbia because they knew the Expedition would soon reach the prairies where game was plentiful.

In order to explore more of the terrain the Captains had decided to separate the expedition into three groups. The first group led by Lewis would head north to the Medicine River, and then follow that river to the Great Falls. He would then split his group taking part of it with him north to explore the upper Marias River. The remainder would wait at the Great Falls to meet Ordway coming down the Missouri. Ordway was going to leave with Clark and go to the head of the Jefferson River then bring the canoes they had stored down the Missouri to the Great Falls. After portaging the falls, he would continue down the Missouri and meet Lewis at the Marias. Clark's group was going to go east from the Jefferson to the Yellowstone and meet the whole group where the Yellowstone dumps into the Missouri.

Lewis took Sgt Gass, Drouillard, Joseph and Reuben Field, Werner, Frazer, Thompson, McNeal and Goodrich. They took seventeen horses to carry the men and the baggage. Clark took all of the rest of the Corps. Ordway and nine men split off at the Jefferson. On July 3rd Lewis and Clark separated with their men and started on their way. Clark had fifty horses and twenty-one men, Sacagawea and her baby, Lewis with seventeen horses, ten men including him, both left camp heading in different directions.

The next day Lewis made it to Rattlesnake Creek, now East Missoula. The Indian guides left them at the Clark's Fork and they were now on their own. However, they had received detailed instructions from the Indians on the route. The heavily traveled trail should be easy to follow. Clark made it thirty-five miles down the Bitterroot River and camped in the area three miles north of present day Hamilton, Montana. Both camps were able to find enough game and grass. On July 4, Clark's group camped early so they could cook fresh deer for a small celebration of Independence Day. Lewis made it to Rock Creek and wrote no word of a celebration or even mentioned it being Independence Day. Both Lewis and Clark continued to their destinations. Lewis followed the Blackfoot River Valley and Clark moved into the Beaverhead drainage. Neither party seemed to be having any problem securing enough game to feed the men. They were moving fast and hard and had their hunters starting out early to secure meat for the troops.

When Lewis crossed the continental divide in late afternoon of July 7, he could see Fort Mountain (Square Butte) in the distance. They were now on the Mississippi drainage side of the mountains and the view was indescribable. The men found sign of buffalo, which they were eager to harvest. Hunting has been successful so the crew and horses were in good condition. Clark moved through the big Hole Valley and camped below Big Hole Pass. The Valley is beautiful and has an abundance of game and beaver. They were now in the drainage of the Mississippi and moving toward the Jefferson River.

When Lewis reached the plains hunting was not as successful as the mountain and foothills areas had been. They made camp where Elk Creek dumps into the Sun Rivers just east of Augusta, Montana. They saw buffalo in the distance and tried to get close enough to kill one. The next day they harvested their first buffalo. They enjoyed the fresh buffalo which was delicious compared to some of the deer and other animals they had survived on. Their encampment was north of the present day Simms, Montana.

Clark made his way to Camp Fortunate at the head of the Jefferson River where they had sunk their canoes and made a cache the year before. The cache held tobacco and the men hurried to dig up the cache to get it. Some of the plants that were collected were mildewed and ruined. They put the canoes in good order and prepared to head down the Jefferson to the Great Falls of the Missouri. The next day they proceeded to where they had cached a canoe on the east side of the Jefferson River opposite the mouth of the Big Hole River two miles North of Twin Bridges.

Lewis proceeded down to White Bear Island where he saw thousands of buffalo, as he did the year before. He instructed the hunters to kill a few buffalo for food and their hides. They made boats from buffalo skins by using willow branches to stretch the hides. These boats called "bull boats" served them well to get their baggage to White Bear Island so they could dig up the cache and be ready to move when Ordway arrived with the canoes. When they dug up the cache, they discovered a lot of damage from moisture. All of the plant samples mildewed and some of the medicine was gone. Lewis stayed at the White Bear Camp another day and set the men to drying meat. McNeil returned to camp late and recounted his adventures with a white bear. He had to sit in a tree for three hours until the bear left. During the fracas, he bent his gun barrel. He was lucky to be alive as he hit the bear with the gun to stun it enough to enable him to climb up a tree.

Clark was moving along rapidly, as the river current was swift. He had Sgt Pryor and party go ahead with the horses to hunt. Clark instructed them to wait where the Madison River joined the Jefferson River. Pryor had killed enough food for the night. Clark loaded most of the meat into the canoes and sent Sgt Ordway, Colter, Collins, Cruzatte, Howard, LePage, Willard, Weiser and Potts on their way down the Missouri. Clark and his party now consisting of Bratton, Sacagawea, Charbeneau, Pomp, Gibson, Hall, LaBiche, Pryor, Shannon, Sheilds, York and Windsor had forty-nine horses and a colt. They proceed up the Gallitan toward the Yellowstone. Sacagawea pointed out the pass her people used to cross the mountains in front of them. They camped that night at the present site of Fort Ellis east of Bozeman. The hunters were very successful in securing game. The next day Clark crossed Bozeman Pass and camped near present day Livingston, Montana at the Yellowstone River.

On July 16th Lewis began his trip to explore the Marias. With him were J. Field, R. Field, and Drouillard. They visited the Great Falls of the Missouri and camped over night. From there he headed north to find the Marias River. They camped along the Teton River, which travels in an easterly direction between the

Marias and the Missouri. The next day they headed up the Marias toward the mountains. Lewis described the plains as rolling endless short prairies on heavy clay soil. The river bottoms produced plenty of game for the four men. With a small party of four, it is not hard to find enough food to feed everyone. The party continued and camped at the entrance of Cut Bank Creek. The Marias River was now obviously not a passage through the mountains so he decided to stop following it. Lewis planned to stay at this camp for several days so he could take celestial observations to fix exactly where they were. For the first time on this trip, Lewis said that buffalo and other game were becoming scarce and their provisions were running low. He sensed that Indians could be in the area as the game was very illusive.

Clark proceeded down the Yellowstone and made thirty-eight miles. Game was plentiful and they ate hearty. He continued looking for trees large enough to make canoes. He finally found two trees that would make small canoes in the area south of current day Park City. Clark could see signal fires of Indians along the hills above the valley of the Yellowstone and kept an open eye for Indians that might approach. Gibson was thrown from a horse that stepped in a hole while chasing a buffalo. He was in much pain while riding which increased the need for canoes. They stopped and constructed two canoes, and with plenty of game in the area, they rested the horse's feet. Some of the men dressed hides while others dried meat.

Ordway had made his way down the Missouri to the Gates of the Mountains. They continued down the river finding plenty of game as they went. On July 19, they arrived at White Bear Island where part of Lewis' party was waiting for them.

On July 23, while Lewis was trying to make observations of latitude and longitude he sent the hunters to find some meat. They were not successful and once again, the men found themselves without food. They had a tainted piece of meat, from which they rendered the fat. Lewis wrote, "we now rendered the grease from our tainted meat and made some mush of cows with a part of it, reserving as much meal of cows and grease as would afford us one more meal tomorrow." **(1)** They evidently still had a supply of roots in their packs, which make great emergency nourishment. Lewis was still trying to get observations of this location. He sent the men to hunt to the south and they were successful, returning to camp with meat. It rained most of the day and Lewis' was unable to make more observations.

Clark's party finished the canoes and got them ready to float down the Yellowstone. He had them lashed together as they were small and quite unstable separately. The Indians stole twenty-four horses a few nights before. Labich found the tracks on the 23rd and could tell that they headed them across the prairie hurriedly.

Clark ordered Pyror, Shannon and Windsor to head overland with the remaining horses and take a letter to H. Haney who was possibly at the Mandan Village. They were to meet Clark at the mouth of the Big Horn later in the day before they continued. The letter explained that the Captains wished to take some Indian Chiefs back to the Seat of the Government to let them see the Indian resources and talk of trade. They were to leave the horses with the Mandan's for Clark to pickup when he arrived. They killed some buffalo and had a supply of dried meat, some of which the wolves had scavenged.

The next day they proceeded down the river to an area southwest of Billings, Montana. They made camp and on the next day proceeded to a mound of sand stone along the river. Clark called this Pompey's Pillar. Here he carved his name and date in the sand stone, which remains there today. As they proceeded down river, they saw immense herds of buffalo and other game all along the way. The buffalo were so troublesome that the men shot their guns over the herds to scare them away from camp.

Ordway's party was busy at White Bear Island. They made wagons to transport the canoes and were busy moving them with horses to the lower portage camp. It seemed wherever they would stop, finding dinner was always near the camp.

On July 27, Lewis gave up on getting observations at Camp Disappointment and headed south toward the Two Medicine River. In late afternoon, he noticed thirty horses and Indians at a distance of about a mile. He was sure these were probably Minnetaries and would probably try to rob them of their horses. Lewis approached very carefully and eventually dismounted and approached in peace. There were eight Indians. They went with

Lewis and made a camp at the Two Medicine River. The Indians stayed the night with Lewis's party of four. Lewis and the men were very careful and guarded their property all night but just around sun up the Indians decided to steal their guns and horses. I cannot describe the situation better than the writings directly recorded in Lewis's journal.

He writes, "This morning at day light the indians got up and crouded around the fire, J. Fields who was on post had carelessly laid his gun down behid him near where his brother was sleeping, one of the indians the fellow to whom I had given the medal last evening sliped behind him and took his gun and that of his brothers unperceived by him, at the same instant two others advanced and seized the guns of Drouillard and myself, J. Fields seing this turned about to look for his gun and saw the fellow just runing off with her and his brothers he called to his brother who instantly jumped up and pursued the indian with him whom they overtook at the distance of 50 or 60 paces from the camp sized their guns and rested them from him and R Fields as he seized his gun stabed the indian to the heart with his knife the fellow ran about 15 steps and fell dead; of this I did not know untill afterwards, having recovered their guns they ran back instantly to the camp; Drouillard who was awake saw the indian take hold of his gun and instantly jumped up and sized her and rested her from him but the indian still retained his pouch, his jumping up and crying damn you let go my gun awakened me I jumped up and asked what was the matter which I quickly learned when I saw Drouillard in a scuffle with the indian for his gun. I reached to seize my gun but found her gone, I then drew a pistol from my holster and terning myself about saw the indian making off with my gun I ran at him with my pistol and bid him lay down my gun which he was in the act of doing when the Fieldses returned and drew up their guns to shoot him which I forbid as he did not appear to be about to make any resistance or commit any offensive act, he droped the gun and walked slowly off, I picked her up instantly, Drouillard having about this time recovered his gun and pouch asked me if he might not kill the fellow which I also forbid as the indian did not appear to wish to kill us, as soon as they found us all in possession of our arms they ran and

indeavored to drive off all the horses I now hollowed to the men and told them to fire on them if they attempted to drive off our horses, they accordingly pursued the main party who were driving the horses up the river and I pursued the man who had taken my gun who with another was driving off a part of the horses which were to the left of the camp, I pursued them so closely that they could not take twelve of their own horses but continued to drive one of mine with some others; at the distance of three hundred paces they entered one of those steep niches in the bluff with the horses before them   being nearly out of breath I could pursue no further, I called to them as I had done several times before that I would shoot them if they did not give me my horse and raised my gun, one of them jumped behind a rock and spoke to the other who turned around and stoped at the distance of 30 steps from me and I shot him through the belly, he fell to his knees and on his wright elbow from which position he partly raised himself up and fired at me, and turning himself about crawled in behind a rock which was a few feet from him.   he overshot me, being bearheaded I felt the wind of his bullet very distinctly._ not having my shotpouch I could not reload my peice and as there were two of them behind good shelters from me I did not think it prudent to rush on them with my pistol which had I discharged I had not the means of reloading untill I reached camp; I therefore returned leasurely towards camp, on my way I met with Drouillard who having heard the report of the guns had returned in surch of me and left the Fieldes to pursue the indians, I desired him to haisten to the camp with me and assist in catching as many of the indian horses as were necessary and to call to the Fieldes if he could make them hear to come back that we still had a sufficient number of horses, this he did but they were too far to hear him. We reached the camp and began to catch the horses and saddle them and put on the packs"
(2)

  Now they had trouble! The Indians were driving off their horses. The Field brothers pursued the Indians and Lewis got between the Indians and their horses. Lewis called out to Drouillard that with the Indian horses they had were plenty to proceed. They began putting their packs on the horses. The Fields returned with four of the horses and helped continue packing with much haste. Lewis knew that they had to ride day and night to avoid another conflict. They packed the Indians' buffalo meat as they had very little left of their own supply. They also used some of the Indian horses since they were in better condition than the ones they had. They left the extra horses. Before they left the scene, Lewis put the medallion that he had given one of the chiefs the night before, around the neck of the dead Indian so that the tribe would know who they were. He had also given them an American Flag. They had left it with some of their other baggage so Lewis retrieved it and a couple of the items of the Indians abandoned baggage then they were on their way.

Lewis set his course across the plains for about sixty-three miles before stopping to rest their horses and took dinner. They passed the area around Conrad, Montana in a bearing toward the Marias River where it meets with the Missouri. There they would meet Ordway with the canoes and continue down the Missouri by canoe. They rode through the night until about 2:00 A.M.; riding on the plains rather than next to the river was faster as the many crossings of the meandering river and terrain was slowing them down.

While Lewis, Drouillard and the Fields brothers were hastily making their way to the mouth of the Marias River. Ordway was casting off the pirogue and the canoes on the Missouri on his way down to the mouth of the Marias. He had Gass and Willard on the north side of the Missouri with the four horses carrying the dried meat. Ordway had spent the night near Fort Benton and Lewis had spent the night on the Teton River near Fort Benton. Unknown to them they were only a few miles from one another. On July 28, both parties headed out shortly after sunrise. About 9:00 A.M., Lewis heard the sound of gunfire over the ridge to the south. They rode to the top of the ridge and could see the Missouri and Ordway and his men killing buffalo. Lewis rode to the group and explained their situation. They loaded their baggage in the canoes, turned loose the horses, threw the packsaddles in the river and joined the canoes. They floated to their cache at the Marias River and retrieved gunpowder, salt, flour and pork that was in good order. However, some of the hides and collections molded as well as the parched meal. All of the baggage was retrieved accept a few traps. While they were working on the cache, Gass and Willard arrived by horseback with the dried meat. All of the articles were loaded in the canoes and white pirogue. The red pirogue was not in any condition to make the trip. They stripped her of all the metal objects and proceeded on down the river with the white pirogue and five canoes.

They continued down the river until late in the day before encamping in the area of Crow Coulee about fifteen miles below the mouth of the Marias River. The hunters immediately killed buffalo for dinner. The meat was roasted over the fire. Lewis's party covered more than 160 miles in two days. A miraculous trip ending almost at the exact time the split party of Ordway reached the mouth of the Marias.

Clark encountered very little trouble floating down the Yellowstone. Elk and buffalo are in great abundance providing plenty of fresh meat every day. Clark commented that the elk are so tame that they do not move when hunters approach within thirty yards of them. The party is moving about seventy miles a day.

Both Clark and Lewis are moving at an astounding speed as the mid summer rains have raised the rivers and they are flowing about five to seven miles per hour. They spend little time on shore, there is so much game found along the shores that it does not take long to procure enough for food and hides. Good clear drinking water is harder to find than meat. They continued downriver and by July 31, Lewis is at the beginning of today's Fort Peck Reservoir while Clark camped about seven miles southwest of Glendive.

Lewis was becoming concerned that some of the mountain sheep hides and horns they had harvested in the White Cliffs area were beginning to spoil. He had the crew pull in to an area called Horseshoe Point, which is about fifteen miles below the mouth of the Musselshell River. An old Indian camp there provided quick shelter. As they were arranging camp, a large grizzly bear approached the men on her hind legs. They all proceeded to empty their guns on her. Earlier in the day, they had killed another bear and had brought it down to this location in the canoe. The grizzly is a great source oil or grease and the meat is fatty.

Lewis had the bear meat cut from the bear and the fat rendered. He described the fat, "the oil of this bear is much harder than that of the black bear being nearly as much so as the lard of a hog." **(3)** They extracted several gallons of grease from the bears. Now they can have fried meat, which is much tastier and faster to cook. Lewis planned to stay a day or so and instructed the men to dry some meat.

At the mouth of Thirteen Mile Creek, which is about fifteen miles above Glendive, Montana, Clark encountered a large herd of buffalo crossing the river in front of them. They had to wait about forty-five minutes for the herd to cross before they could proceed. The hunters killed four fat cow buffalo and removed as much fat from them as possible. It is interesting to point out that on the same day and miles apart both parties secured a supply of grease.

Lewis writes on August 3rd, "we did not halt today to cook and dine as usual having directed that in future the party should cook as much meat in the evening after encamping as would be sufficient to serve them the next day; by this means we forward our journey at least 12 or 15 miles Pr. Day." **(4)** Clark reached the Missouri where he was going to spend a day to dry their gear and wait for Lewis. However, the mosquitoes were so bad that they could not stand the torment. Pomp's face swollen from insect bites left him very miserable. Clark wrote a note to Lewis and stuck it on a stick telling him they were moving downstream where the mosquitoes were less troublesome. They packed up and moved down stream.

By August 7, Lewis finally arrived at the confluence of the Yellowstone and found Clark's letter. In order to get to his camp they continued until late evening but did not find the encampment. Early on August 8, Lewis not finding Clark's camp was not sure that they did not pass them on the river, Lewis stopped to set up camp. The pirogue and one of the canoes needed repair and the men needed a break to fix their clothing and, of course, collect a little meat. Lewis knew that they soon would be leaving the plentiful bounty of game and would need a supply of dried meat.

On August 11, Lewis proceeded on to find Clark. They saw a herd of elk on shore and pulled over to hunt. He and Cruzatte climbed out and began to stalk some elk. Lewis shot one and reloaded to fire at a second when a bullet passed his upper thigh below the buttocks. Cruzatte had shot Lewis by mistake. He would not admit to it saying he was shooting at an elk. Lewis recovered the ball and determined it belonged to a short rifle like the one Cruzate was using. The injury was not fatal. Lewis dressed the wound and suffered with pain for weeks. Cruzate, not having the best vision, took him to be an elk.

They proceeded on and found one of Clark's abandoned camps. There was a letter informing him that they are ahead and a note saying that the Indians had stolen some of their horses. On the way downriver, now in bullboats and frantically trying to catch up with Clark, Pryor took a letter that Clark left for Lewis.

As Lewis continued down the Missouri, they met mountain men, Dickson and Hancock, who were heading up the river to trap. These men informed Lewis that Clark was about one day ahead of them. Dickson and Hancock decided to return to Mandan with the party in hopes that they could acquire a Frenchman or someone that knew the area to accompany them on their trip back upriver, and of course hear more about any locations the Corps thought would be good trapping grounds.

They awoke early the twelfth and proceeded down stream to overtake Clark. At about 1:00 P.M., Clark's camp was in site. Lewis' buttock was extremely painful, but everyone else was healthy and extremely happy to see each other. Clark told Lewis that Pryor lost the horses near Pompey's Pillar, chased the Indians for several miles and finally gave up. Pryor proceeded to build bullboats from buffalo hides then preceded down the river very successfully without taking on much water.

A few days earlier, Clark's men dug breadroot for dinner. Clark wrote, "the men dug great parcel of the root which the Nativs call Hankee_ and the engagees the white apple which they boiled and made use of with their meat. This is a large insipid root and very tasteless. the nativs use this root after it is dry and pounded in their Seup." **(5)** They described their experiences with each other and visited as hunters do about their camps today. Lewis decided not write in his journal for a while until his wound heals, so Clark continues the journals for both. They proceeded on the next day and covered eighty-six miles.

The next day, August 14, the combined Corps of Discovery passed the first villages of the Hidatsa and Mandan people. Some of the Chiefs recognized them and being happy to see them, showered them with gifts of corn, beans and squash. This was a real feast for the Expedition since they have not had any type of vegetable except prairie roots for months. The fresh corn harvested at this time of year was a real treat. Clark called a summit of all the Chiefs in the area to see if he could get a few of the chiefs to go with them to visit the great father the president of the United States, present themselves, receive gifts and talk of trade and peace. It was a tough sell because they were at war with the Sioux Indians they would have to pass through along the way. A couple of Chiefs was considering the offer, but could not make a decision.

When he returned to camp that night John Colter requested his release from the Expedition so that he could join Hancock and Dickson and go back to the Yellowstone and trap. The Captains gave him permission as long as no one else had a request to do the same. They supplied Colter with some lead, powder and other articles that the Expedition would not need. They also settled with Charbonneau and discharged him. Sacagawea and Pomp, who is now nineteen months old, also stayed at the village, but Clark made a promise to Charbonneau that Pomp could come to the United States and get a proper education. Charbonneau said that he might take him up on that offer after Pomp is a little older.

Clark was successful in getting one of the chiefs, "The Big White Man Chief" of the Mandan Village, his wife and child, and the interpreter Jessomme and his wife and two children in making the trip. Clark tied together two of the larger canoes and made a platform that could accommodate a larger load. The Corps then moved downriver to the area of Fort Mandan where they had spent the winter. Here they made camp and were able to kill several elk and deer. The meat was loaded on board for future use. As the Expedition made its way downriver, there were no confrontations with the Sioux and they approached the Ricaras and the Cheyenne. They smoked with the tribes and Clark again tried to get someone to come to talk with the President of the United States.

It was August 26; they were in the area of the Teton Sioux confrontation when they were heading up the Missouri in 1804. They were on guard watching for any intrusions. They traveled hard all day without stopping in hopes to get far enough away from the Sioux Villages to prevent any problems. They camped at dark having covered 60 miles that day.

The hunters were not having problems securing enough to eat and with the store of corn, and vegetables from the Mandan's, very few complaints of being hungry were voiced. On August 29, Clark walked on shore and could view thousands of buffalo grazing on the prairie. He was surprised to see such large herds since they were close to several tribes of the Indians.

Captain Lewis' injury was healing slowly. He could now move about and was feeling better by the day. The next day, a band of the Teton Sioux approached them and hostilities cast in both directions. Clark threatened them to stay away or they would kill them all. They were a warring tribe and made hostilities toward white traders that approached their villages. Clark moved to an island in the middle of the river as the Indians watched in the distance. No confrontations arose but they were on guard all night.

As they continued moving downstream on 1 September, they encountered a group of Indians they thought were Teton Sioux, but turned out to be Yankton Sioux. They smoked with them and gave some medals and ribbon. The Corps camped that night near Yankton, South Dakota. The wind blew violently all day. They continued in late afternoon after taking some buffalo meat on board for each canoe. Clark also mentioned the taking of two turkeys. Again, this is Clark's favorite fowl. He does not say a word about them cooking it, but I am sure they did and as fast as they were traveling, they probably boiled it in oil before they retired for the day.

On Sept 3rd they met with James Aird, a trader along the river, who related some political news from United States. The men received tobacco and a keg of flour in exchange for some of the fresh corn they had on board from the Mandans. This was welcomed since the little flour remained Marias River Cache. It probably was stale and may have been full of bugs. They did not mention using any of it. A cup of flour was issued to each man . They now could probably make some fry bread.

They were moving rapidly along the river, but took time to visit Sgt Floyd's grave near Homer, Nebraska. The Indians found Floyds grave and started to dig thinking it was a cache. When they found it was a burial site, left it partially uncovered. The men recovered the grave then headed back down the river.

They continued meeting traders heading up the river. They purchased a gallon of whiskey from Mr. Choteau, but he would not accept any payment. They did trade some of their leather goods for some linen shirts and some hats. The men were drooling for a taste of the whiskey, as it was the first they have had since July 4, 1805. Each man enjoyed one dram. They continued onward and were lucky enough to find some elk to kill for dinner as well as three fine catfish. They moved on covering seventy-eight miles before camping near the mouth of the Platte River.

The Indians were getting restless and homesick on the voyage and the children would cry because of the uncertainty of the destination. The men on the other hand are pushing hard to advance as their anxieties of being home are building. They are making about seventy miles a day and will soon be back in civilization. Every day that passes, more traffic and traders are heading up the Missouri to trade with the natives. On Sept 12th they met Mr. McLanen who was traveling up the river. He gave them more whiskey and three bottles of wine for the captains. He also gave Clark a piece of chocolate, which Clark used to make hot chocolate to drink the next morning. As they continued, they begin to collect plenty to eat and drink. They were anxiously awaiting the ending of this journey. Occasionally they would stop in this area and the men would collect "pawpaws" or *Asimina triloba*, a fleshy fruit about six inches long and two inches thick that hang down like bananas and have a texture of sweet potatoes. It was one of their favorite river fruits.

By September 20, they were only days away from St Louis. Gass quit writing in his journal and game collecting ceased. They did not have game on board and were subsisting on Paw Paws. The men did not seem to care as they said they could subsist on Paw Paws until they arrived. Therefore, they pushed on. When they reached the small village of LaCharette the men were given money to purchase some whiskey and food. They knew they were very close to St. Charles and could not wait to celebrate, which they did. Clark does not make a big deal of it in his journal. All he said the next day was that we collected our men and headed out.

At four o'clock in the afternoon on September 21, 1806, they arrived at St Charles. It was a Sunday afternoon, and not a lot of action on the shoreline until the men of the Corps of Discovery started shooting their blunderbuss and rifles in celebration. The shoreline began to fill with people. The residents thought that the Lewis and Clark Expedition had probably perished in the western wilderness on the wrong edge of survival.

The next stop was St Louis. They arrived in St Louis at 12:00 noon on Sept 23, 1806 and the celebration that followed was jubilant. Then Ordway writes in his last entry. "drew out the canoes then the party all considerable much rejoiced that we have the Expedition Completed and now we look for boarding in Town and wait for our Settlement and then we entend to return to our native homes to See our parents once more as we have been So long from them— finis." **(6)**

Part III
Collecting and Evaluating Information

One who reads the Expedition's journals will soon become familiar with their frequent killing of buffalo, elk, deer, antelope and birds for food. It is easy to assume they are eating vast quantities of meat and little else. But a closer examination tells a much different story. The journals record at least 94 different foods the Expedition ate during their 28 month journey. They harvested 46 of these foods from the land as they passed through.

The problem is the journal keepers didn't record in detail everything they may have for a meal. Just like today we may say we had a steak dinner, but we don't say we also had mashed potatoes with sour cream and butter, tossed salad, dinner rolls, cherry cobbler and ice cream, etc. They seldom mention when they ate most of the rations they took, but, except for some they cached, these supplies were exhausted a year before the journey ended.

Chapter 14
Diet during the journey

Almost as soon as the Corps of Discovery left their winter camp near St. Louis on the journey up the Missouri, we see a change in their eating habits as well as their diet. The Captains' detachment orders for May 26, 1804 specifies that cooking will only be done in the evening. Other meals will be leftovers from the night before. Although not stated, it is presumed that while at Camp Dubois all meals were cooked since that was typical for the army in garrison. This order also specified what provisions would be issued each day. From examining that order and later statements in the journals, the Captains knew they had to conserve the food they brought with them to make it last as long as possible. So the Corps depended more heavily on the fresh meat and other foods they could get along the way. As they progressed up the Missouri more wild fruits and berries became part of the men's diet.

During the summer and early fall of 1804 the Corps of Discovery ate well, mixing a variety of game animals, fish and wild edible plants such as fruits and berries with some of the cornmeal, flour, lard and occasionally pork they had brought with them. By the time they reached the Mandan Villages in late October the changing seasons forced a diet that had a much greater reliance on meat from the deer, elk and buffalo the hunters got and much less on edible plants along their route.

During the winter of 1804-1805 at Fort Mandan, corn from the Indians in exchange for blacksmith work became more important as fresh meat became scarcer. As the fall harvest of wild fruits and berries gave out, along with the availability of game animals, the Corps diet changed to one of basic subsistence consisting of what little meat, corn and squash they could get. The Mandan Chief Big White summed up the situation best by saying the Indians and the Corps must share their foods when they had any, "If we eat you shall eat, but if we starve you too shall starve."

At Fort Mandan food was frequently shared between the Corps of Discovery and their Indian neighbors. Both Indian and Expedition visitors almost always gave gifts of food to their hosts during the many visits between the two groups. Joint hunting trips were the order of the day. It was common to see several Indians join several of the Expedition hunters in a united effort to get meat for everyone. Indian horses and Expedition guns combined for much greater success than either group had alone in this austere winter landscape and scarce game.

Spring of 1805 brought a new abundance of food for everyone, primarily in fresh meat from the deer and buffalo that returned to the plains. As the Corps of Discovery continued their journey west toward the Rockies they ate well. Clark commented the plains were like a commissary where they could select almost at their leisure what they wanted to eat. He was referring to the variety of animals available for meat, but the prairies also provided a variety of plants for their diet. Sacajawea often dug roots when the Corps stopped to dry baggage or wait for the wind to subside. With vast herds of buffalo, elk, deer and antelope as well as many bighorn sheep and beaver so readily available both Captains cautioned to only take what was needed.

The next big change in diet was made as the Corps made it way through the Rocky Mountains. The Captains carefully expended the provisions they brought from Camp Dubois to make them last as long as possible. They knew when they reached the Rockies they would not have the readily available supply of meat that they enjoyed on the plains so the corn, flour and pork would be needed. They also had to save some of these staples for their return trip; several barrels of these provisions were cached at the Marias River.

The Expedition laid in a good supply of "jerked" meat and continued actively hunting, but by the time they were in the heart of "those terrible mountains" game was almost nonexistent, with only an occasional meal from fresh meat. The last few days in the Rockies were a near starvation ordeal. All the fresh meat, jerked meat, provisions from St. Louis and even the portable soup was gone (portable soup gave the Expedition six meals in the Rockies). Survival had meant eating three of their horses.

Diet during these days in the Rockies was bare subsistence, eating whatever precious little food came available. They never really had enough to satisfy their hunger; instead it was just enough to survive until something else could be found to eat. Early snows added to the problems of finding food. The list of what was eaten included mainly meats since the snow made the roots and berries impossible to find.

Arrival at the Nez Perce camps on the Weippe Prairies of Idaho meant the Expedition again had food. Their diets changed to be much like that of the Indians in the area; dried salmon and camas roots with an occasional deer. This was the most severe diet change the Expedition faced and the results were immediate and telling. Camas root is an emetic as well as a purgative until the system adjusts to it. **(1)** The men had almost nothing in their systems so they felt the fullest affects of the camas roots' medicinal values. Within hours everyone was cleaning their systems from both ends. It took several days before they adjusted to their new diet. The only relief from the effects of the roots the Expedition got during their stay in the area was by eating horse, dogs or the very few deer the hunters could find. Everyone's systems had adjusted to the camas roots sufficiently to be "tolerably well" by the time the canoes were completed and the Expedition set off from their Canoe Camp, down the Clearwater River. The whole process of adjusting to this strange new diet was made worse by the entire Expedition getting extremely sick when they first arrived at the land of the Nez Perce.

From the camp with the Nez Perce until the Expedition reached the Pacific Ocean, their diet remained dried salmon, camas root and dog meat. This period of time must have been particularly difficult for Captain Clark since he did not like or eat dog meat. He commented, somewhat wistfully, that Lewis would eat anything if it had enough salt on it, but he wasn't much concerned about salt and could never manage to eat the dog meat.

Once construction of the winter camp at Fort Clatsop was completed, the Expedition was able to turn their attentions to finding their own food rather than trading with the Indians of the area. Their diets once again became one with more meat in it as the hunters managed to regularly find a limited amount of elk to add to the dried salmon and camas root. The men responded to the addition of meat with increased strength, but Captain Clark vividly illustrated the fact that their strength was not nearly what it had been while they enjoyed the plentiful meat harvest of the plains. When he led a group to get some of the blubber from a whale that had been washed up on the beach, they met a group of Indian women climbing a steep hill. Clark took the load from a woman who had slipped, but to his surprise he was barely able to lift the 100 pounds she carried.

The Expedition ran into a new limitation placed on their diet that winter. Although the Indians had a variety of roots, berries and fish and they were willing to trade with the Corps of Discovery, the prices they wanted for these food items was very high. Since the Expedition had very little to trade with, most of the time no deal could be reached. Consequently the men had to depend upon their rifles to supply enough elk meat.

The abilities of the Expedition hunters were definitely tested during the winter, but they apparently passed the test. Clark, somewhat wistfully, noted in his journal as they bade farewell to Fort Clatsop that they had eaten three meals every day they lived there.

The Corps of Discovery's diet remained principally dried fish, roots and elk or deer meat during their trip from Fort Clatsop up the Columbia then to the Nez Perce and over the Rockies to Travelers Rest; basically the same diet they had during their winter at Clatsop. There were variations in the amount of meat and at times instead of elk or deer it was fresh fish or dog. While they were camped with the Nez Perce waiting for snow in the Rockies to melt, the hunters got several bear which not only was used for their meat, but the bear oil that they used in preparing their dried roots.

When the Expedition left Travelers Rest, Clark headed towards the Yellowstone and Lewis overland to the Great Falls. Their diets changed to be primarily meat—deer, buffalo or elk. They had eaten most of the roots and other dried foods obtained from trade with the Indians and fresh plant foods were not yet ready. The provisions they had cached were at the mouth of the Marias which Clark completely bypassed on his Yellowstone route and Lewis didn't reach until July 28th. However, they varied their meat diet by including geese, beaver, rabbit and catfish. Clark drew the lucky route since his entire trip from the time he left Travelers Rest until the two groups reunited in western North Dakota ran through an area thickly populated with game. They did not want for food at any time.

Lewis' easterly trip to the Great Falls was also through land well populated with game, so his party ate well on much the same foods that Clark's group had. At the Falls Lewis split his group, taking three others north with him to the far reaches of the Marias River. Those that remained at the Great Falls continued to eat well, but Lewis ran out of food in an area that had been heavily hunted by Indians very recently. Consequently his group spent several days with very little food. A small quantity of roots, the last of the supply from the Nez Perce, one small trout, several pigeons and a small piece of buffalo meat.

Lewis' party returned to the Missouri River and recovered the corn, flour, pork and salt from their cache at the mouth of the Marias. Since they were back in the land of plentiful game, they ate well the rest of their trip downriver to where they were reunited with Clark.

As the journey downriver continued, the Expedition left the lands of plentiful game animals and entered an area of a much reduced meat supply. The area was home to a larger population of Indians who also depended upon the game for their meat. Diets now were much more dependent upon corn the Expedition got from the Mandans through trade and gifts, wild fruit and berries, and other birds and animals such as turkey, pelican, prairie dogs and a few rabbits. The men had enough to eat to curb their hunger, but not to fully satisfy their appetites.

By early September the Expedition was about two weeks travel away from St. Louis. Their diets again changed as they started meeting traders coming upriver. From these traders the Expedition got a limited amount of flour, biscuits, chocolate, sugar, beef and pork. That, along with plums and papaws, satisfied the men who were more interested in reaching civilization that what they were eating. They had stopped hunting altogether, preferring to use the time and effort getting downriver and home. They had entered an area that was fairly heavily populated by various native tribes who depended upon the same game animals the Expedition sought so hunting required a greater amount of time than the Corps of Discovery could afford.

The journals do give a fairly complete list of the foods the Corps of Discovery ate, particularly once the journey got under way. Food usage information during the winter at Camp Dubois is much less complete. What is written about the foods they ate lead to the logical conclusion that they ate typical frontier/military diet. See appendix III for the complete list.

Chapter 15
Assembling the data

A careful examination of the Expedition's journals and the collection of related documents show a short list of cooking utensils along with the staples taken from St. Louis. True to Jefferson's requirements a military unit was formed and took most of their food from the land and through trading with the people along their route. They reserved most of the rations they brought from Camp Dubois for those times to come when food from the land was not sufficient to feed the Corps of Discovery. Those documents also tell us some of the cooking techniques used. But those documents don't provide the complete answer to the initial questions. What did they eat, how was it prepared and what equipment was used to cook with?

When reviewing equipment the Corps took along for preparing their food we will look at kettles, dutch ovens, coffee grinders, personal equipment, cooking utensils, frying pans and fire irons. We know from the records they took brass kettles, corn mills and they cached a dutch oven. They had to have some means of frying and roasting their foods since references are made to those types of cooking and something to suspend their kettles from when hanging them over the fire. We must consider what they ate and drank from and how they prepared their coffee (The records show the Expedition took 50 lbs. of coffee and Clark mentions that he had a cup on July 19, 1804 and on June 25, 1805 which was the first coffee he had since winter at Mandan).

Some Hints to Consider

The records of the Expedition as recorded by the journal keepers are very interesting reading. Because they contain so much detail it seems that every time they are re-read something new is discovered. Although they recorded very little detail about their cooking habits there are some key bits of information to be found if the journals are read carefully. The tidbits that are uncovered don't tell us a great deal by themselves, but when corollary sources are added we start to see some more substantial pieces of information.

The men of the Corps of Discovery were not ones to waste a lot of time with food and cooking. Most of them were military enlisted men and were accustomed to a very simple, bland diet. They didn't care much what they ate, just as long as they could eat something. They knew that they would only get food that could be prepared in a very quick, simple manner since the cooks were not chefs. They were only cooks because they had been assigned the duty. During much of their traveling time the Expedition would stop in the late afternoon long enough for their evening meal then continue their trek until dark. The maximum amount of time spent at the stop was two or three hours to set up and cook the food, eat, maybe take care of other pressing matters, then pack up and continue on.

There are some hints given in the journals that could be saying the Expedition had more cooking equipment than the list of items Lewis purchased in Philadelphia. The order written by the Captains on April 1, 1804 that first established the messes says that the camp kettles and other pubic utensils for cooking would be divided into the messes. Later on April 13th Clark returns to Camp Dubois from St. Louis with sundry articles for our voyage. The journal for the Expedition's time at Camp Dubois was extremely brief, with daily entries frequently limited to a single sentence. Since Clark, the only journal keeper at Camp Dubois, does not elaborate on the sundry articles we don't know what he really had. Then on August 29, 1804 the Captains gave a band of Sioux Indians they were going to council with six kettles to cook their meat in. They also gave a kettle to Mr. Dorion, a trader who lived with these Indians. Add to this the fact the Expedition lost cooking utensils on two occasions when they had near mishaps negotiating rapids in the river with their dugout canoes. Ordway wrote in his journal that the canoe he was in on October 14, 1805 was one of these incidents and they lost one small brass kettle.

A final journal excerpt that strongly suggests other cooking utensils is when Clark notes the men who were sent to set up the salt works near Fort Clatsop took five of the largest kettles with them. The list of Lewis' purchases in Philadelphia shows 14 brass kettles. If six were given to Indians at Council Bluff and one was given to Mr. Dorion, then the saltmakers took five more, and at least one kettle was lost in a canoe mishap, the main party of 28 at Fort Clatsop would be left with at most one small kettle for their cooking needs. In Philadelphia Lewis purchased one-five, one-three, two-two and one-one gallon kettle for the Expedition's use so the smallest one left would probably have been the one gallon size. The other eight that he purchased were for presents and trade with Indians; their sizes were not specified. We have learned from our field experiments that one kettle is adequate to feed that size group, but it needs to be larger than one gallon. We used a five gallon kettle to feed thirty people a basic meat, poultry, fish and bread meal. In retrospect we could have probably used as small as a three-gallon kettle to do the same job.

These hints, when taken together, become compelling evidence that Lewis' Philadelphia purchases were only part of the cooking equipment the Corps of Discovery had when they left Camp Dubois on their journey up the Missouri. The challenge is to determine what additional items they had and where they came from.

The Military Unit Theory

A theory frequently used to fill in the gaps of information regarding cooking methods and equipment is the Military Unit theory. **(1)** This theory says that since the Corps of Discovery was a military unit they would follow Army orders and policies. Therefore to determine how foods were prepared and what equipment was used turn to military standards of that time.

Jefferson's request to Congress for funding an exploration of the northwest called for a military unit consisting of one officer and a dozen enlisted men. A civilian interpreter was approved by the War Department and at Lewis' request an additional officer was added.

When the Corps of Discovery was formed at Camp Dubois all volunteers that were not already in the military were enlisted except George Drouillard, the civilian interpreter Lewis hired. The French engages hired for the trip upriver as far as the Mandan Villages were all civilians and did not enlist. The crew that brought the keelboat back downriver to St. Louis the next spring was military men. Two other men hired at the Mandan Villages to replace the ones who deserted were enlisted into the Army. Even though the size of the Expedition nearly tripled from what Jefferson initially specified, the military unit requirement was maintained.

The Military Unit theory says that the added equipment required for the larger force would have come from the public stores at several army forts in the area. That equipment, if in fact any was added, consisted of more brass kettles similar to what Lewis purchased in Philadelphia. He had used military standards for making his purchases. The Expedition would have only taken absolute necessities and taken nothing that was excessively heavy; brass or copper kettles rather than cast iron for example. Carrying heavy pots and kettles or other metal items across the Rocky Mountains would probably have been impossible.

Since army orders specified boiling as the only acceptable method of preparing foods there was no need for roasting or frying equipment. Consequently none was taken—no roasting racks or spits, frying pans or baking ovens. Because of the expense of iron and the weight involved iron tripods for holding kettles over the fire were not taken. Tripods would have been fashioned from green wood at hand when the Expedition made camp for the night.

This theory addressed the question of dutch oven(s) by saying they didn't take cast iron pots or kettles because of their weight and because food was prepared by boiling they would not have been needed. In response to the dutch oven cached at the mouth of the Marias River, this theory responds by noting that dutch ovens were made from several different metals and in many shapes and sizes; a variety of pots and kettles were referred to by that name. Therefore there is not enough information to really know what was cached. It is possible however that a large brass or copper pot of 12 to 20 gallons in size that had been used for rendering animal fat was cached there because although it transported well in the larger boats, it would not fit well in the small dugout canoes that would be used in the smaller rivers they were entering. Such a large, heavy pot could not be successfully carried in Lewis' iron framed boat because it would punch holes in the elk skin boat covering.

Other parts of this theory explain the lack of records showing purchases of plates, spoons, knives and forks by saying at that time standing army orders specified the individuals would furnish their own eating utensils. **(2)** There are many records that show the men of other military units did, in fact, supply these items for themselves. Spoons, spatulas, forks, etc. used for cooking would have been fashioned from wood along the way since iron ones would have added unnecessary weight. Wooden cooking utensils were frequently used on the frontier.

The Military Unit theory looks at what was recorded by Lewis in Philadelphia and accepts that as what was taken. It says if there were other items where is the record showing the items. The items obtained from the forts along the Ohio and Mississippi Rivers were not recorded because Lewis did not pay for them; these items were simply transferred from one military unit to another.

Proponents of the military unit theory insist that everything must be documented to be included. If it isn't written in a primary source then it can't be used. They continue by saying if these limits are not set and adhered to it becomes impossible to distinguish fact from fiction.

Corn Mills and Coffee Grinders

Some foods only mentioned in passing by the journal writers may tell us by implication of other equipment they had that is not on Lewis' short list and point us to other sources to determine the details. One such mention was made by Clark on June 25, 1805 when he says he had a cup of coffee. By implication we deduce they had to have a means of roasting the green coffee beans and something to grind the roasted beans before the cup of coffee could be brewed.

At the time of the Lewis and Clark Expedition virtually all coffee was sold in whole beans that were green and required roasting and grinding before a pot of coffee could be made. **(3)** The green beans kept for a long period of time without loss of taste. The Expedition left Camp Dubois in mid May 1804 with 50 lbs of coffee. They still had some over a year later when Clark made his remark. When coffee was to be made the beans were roasted over the fire, or on the stove, in a pan. After roasting the beans were ground by hand in one of several ways. Since the coffee bean is soft compared to grains and spices that were also ground in the home, the same equipment could be, and usually was, used. The stone roller was used, as was the mortar and pestle.

Primitive coffee mills (grinders) were hand made by blacksmiths in many communities from the late 1700s to the mid 1800s. They were probably first invented and used for grinding spices, but people soon learned they worked well for grinding their coffee beans too. By the end of that period many changes to the coffee mills were made as their manufacture became more centralized and commercial. Small companies sprang up, each with their own unique style of coffee mill, trying to corner the market for home use. As the population grew it became commercially viable to roast and grind the coffee beans before they were sold. Even from the beginning though coffee mills were expensive for the common person, so most used stone rollers or mortar and pestles which were much cheaper.

Reviewing the general history of coffee preparation we see that during the time of the Expedition they had several choices for roasting and grinding their coffee. We know Lewis purchased three corn mills in Philadelphia that would have worked as well for grinding coffee beans as it did for grinding corn. He may have purchased a mortar and pestle to go with his medical supplies which also would have worked for grinding coffee beans. And finally, since Lewis makes reference in his journal to frying fish in a little grease we know they had some means for frying; that same utensil could have been used to roast the coffee beans.

We can probably conclude that the Corps of Discovery took no utensils or equipment specifically to use to roast and grind the coffee beans they took with them. They chose, instead, to use equipment that had already been purchased for other tasks, but they knew from experience would work for preparing their coffee and was commonly used in the frontier homes.

Dutch Ovens

The use of heavy cast iron kettles and pots for cooking has been commonplace in America since the early 1700s and elsewhere throughout the world much earlier. **(4)** So it should not be a surprise to see a dutch oven referred to by the Expedition journal keepers. Although not in Lewis' list of purchases, a dutch oven was included in the list of items cached at the mouth of the Marias River in 1805. If it true that the name "dutch oven" was loosely applied to a wide range of copper, brass and cast iron pots and kettles of various shapes and sizes, what was this "dutch oven" that was only mentioned by two of the journal keepers and only on one occasion? Where did the Expedition get it and when was it used?

Ragsdale's history on dutch ovens continues. Metal cooking pots (which are round-bottomed and larger at the middle than at the top and bottom), and later, kettles (flat-bottomed and straight sides) have undergone continual change since they were first used in the 7th century. Cast iron was the preferred metal for these pots and kettles because of their heating qualities, but weight and cost considerations made brass and copper ones equally popular.

Most of the history on metal cook pots and kettles was not recorded because they were individually made in small shops scattered throughout the known world. Sizes and shapes varied as much as the individuals who made them. A general story of the early days of metal cooking equipment has been handed down over the years verbally and in many books. Very little of the details survive. However, existing records do show that as early as 1650 iron foundries in Lynn, Braintree and New Haven were making cast iron pots while foundries in Saugus and Batsto made dutch ovens. **(5)** They all bragged that their products were lighter weight than ones imported from England or Holland.

Pots were the earliest design to be used. They were hung over the fire and used to cook soups and stews, and for boiling foods. Their height above the fire regulated the cooking heat. As these pots evolved someone flattened the bottom on one to set it on the hearth without tipping and spilling its contents. Eventually the sides were straightened and a tight-fitting lid was added. The story goes that Napoleon did not like these pots because ashes from the coals on the lid fell into the food when the lid was removed. He called for changes to stop ashes from getting into the food. The first flanged lids resulted. However, in 1790s Paul Revere is credited with perfecting the flange design we know today. He also gets credit for putting the bale on the oven and the three legs. (More on this in chapter 16, Myths Exposed)

With the development of a lip on the flanged lid to hold ashes from falling into the food and the addition of legs, these cast iron pots, or dutch ovens, could be used for nearly all kinds of cooking because the food cooked from the top and bottom. Breads especially require this type of cooking. As an alternative the lid could be removed and the dutch oven was then hung over the fire and used as a pot or even as a skillet for frying foods.

The name has several "origins" depending upon who is telling the story, but probably comes from the Dutch process for casting the pots. The Dutch (in Holland) cast brass pots in sand molds. In 1704 Abraham Darby went from England to Holland to see their method. He eventually refined the process and began casting the heavy pots used in cooking and shipped them worldwide. Darby revolutionized the production process somewhat like the assembly line did for manufacturing. Thus he enabled cast iron pot production and widespread ownership to become a profitable business. This major change could easily justify a new product name.

By about 1800 the term "dutch oven" had come to mean a cast iron, flat-bottomed kettle with a tight-fitting lid and three legs on the bottom that could be used for cooking almost any food from baking breads to slow cooking stews, boiling foods, and even for frying in. These kettles ranged in size from 8 to 18 inches in diameter and 4 to 6 inches deep. They had become an ingenious cooking utensil because of their adaptability. Almost anything could be cooked in this one pot.

Late American history (mid to late 1800s) which includes most of the western settlement like the mountain men, the Oregon Trail, and the Mormon migrations, include the dutch oven as a necessity for cooking. It is reported that the people traveling the Oregon Trail would throw out their heavy cook stoves to lighten the load in their wagon, but kept their dutch ovens to do the family's cooking in. **(6)**

Personal Equipment

Most of the men who became members of the Corps of Discovery were already in the army. What might they have brought with them from their old assignments when they joined Lewis and Clark? Standing army orders of that time required the men to provide their own mess gear; plate, cup, knife, fork, and spoon. Consequently, if the Corps of Discovery adopted this general order, it is reasonable that Lewis would make no mention of these on any equipment lists he would have.

The other piece of cooking equipment the military men may have brought with them was tin cooking pots. These 2 ½ gallon tin pots were in widespread use throughout the army and had been since pre-Revolutionary War times. They had become the primary means of preparing meals. Nothing is said about the men of the Corps of Discovery having these tin pots, but there is also no mention of any clothes or equipment of any kind the military men brought with them, although it is generally agreed they brought their uniforms and weapons.

These tin pots (actually they were sheet iron coated with tin) eventually evolved from a much heavier cast iron pot. Camp kettles, as they were known as, were issued one per mess. The following illustrates that fact. "They were made of cast iron and consequently heavy. He was beat out after carrying the kettle most of the night and having no success in persuading another to take it he told his mess mates he could carry the kettle no further. He eventually sat it down in the road and one of the others gave it a shove with his foot and it rolled down against the fence, and that was the last he ever saw of it. When they got through the night's march, they found their mess was not the only one that was rid of their iron bondage." **(7)** "In future the camp kettles are always to be carried by the messes, each soldier of the mess taking it in his turn" **(8)** "In 1778 British army orders were to mark each camp kettle with the company and mess to which it belonged." **(9)**

Frying Pans and Cooking Utensils

Probably anyone who has cooked anything from hotdogs to roasts or stews over a campfire would agree that forks or spoons with long handles are really convenient. From the earliest times of open fire cooking utensils have been used. Long handled forks, spoons, spatulas and ladles made from iron were very commonly used during the seventeenth and eighteenth centuries. Almost every home in the colonies had its assortment of iron utensils to aid the cooks working over the fireplace.

These household items were produced in most of the local foundries scattered throughout the colonies. Iron ore was found in many places, in small deposits. **(10)** Almost any blacksmith could turn that ore into a variety of tools needed in the homes and on the farms. As long as these products were locally produced they were not expensive. It was the transportation costs that made iron and iron products expensive so items that were made only in certain places that had to be shipped to markets were more costly. **(11)**

Cast iron frying pans are another common piece of cooking equipment that needs to be examined when determining if the Corps of Discovery took other cooking equipment beyond the few items listed on Lewis' receipts from Philadelphia. Journal entries during the journey refer to frying some of their foods so we need to figure out what they used to fry in.

Long handled and short handled frying pans made of cast iron are known to have been used as early as 1634 in the colonies. Household inventories of that date have been found by collectors of cast iron cookware **(12)** Records from the furnaces at Lynn and Braintree show that by 1650 in addition to cast iron pots they were making cast iron frying pans. Frying pans continued in common use up to the present day.

Since iron forks, spoons, spatulas and ladles were all very commonly used and normally available or easily made it is quite likely the Corps may have taken some with them. These are collectively referred to as cooking utensils; several journal keepers as well as the Captains' order of 1 April 1804 refer to their cooking utensils.

The idea of frying pans on the trip is a different matter. Frying pans like other cast iron cookware is not easily made by any competent blacksmith. Consequently cost and availability must be considered along with the question, "can we do without it or use something else in its place?" We know from journal passages the Corps took at least one dutch oven. We also know that a dutch oven can be hung over a fire from its bail and used as a good substitute for a frying pan providing it had a flat bottom. Therefore the Corps probably did not take frying pans with them.

Those Mysterious Fire Irons

The group under the command of Captain Clark that had made the trip down the Ohio River landed at what would become Camp Dubois on the bank of the Wood River at about 2 o'clock the afternoon of December 12, 1803. Here they made camp and started construction of the winter quarters for what would become known as the Corps of Discovery.

Clark was the only one who kept an account of the five month stay there; his notes are now called "The Dubois Journal" by Lewis and Clark historians. That journal "consists of twelve loose sheets of different sizes, on which are found not only dated entries but other miscellaneous, undated material; the thrifty captains used the same sheets for a variety of purposes...on many sheets the writing runs in several different directions...it is extremely sketchy and disorganized." **(13)** Consequently only fragments of the events during the winter are recorded. We do know from Lewis' detachment order dated February 20, 1804 that he had the blacksmiths busy making a list of items. **(14)** We don't know what they were or when they would be used, but since they were relieved of guard duty until they were finished producing whatever they were making Clark must have thought these items were important to the success of the expedition. Could they have been making items that would be used later during their journey to the Pacific?

Before we explore the possibilities of what the blacksmiths were making we need to regress a bit. The term fire iron as it is used in this book refers to items used to suspend food over a campfire so the food can cook. There are two kinds to be considered.

First, fire irons include a tripod that is used to suspend a kettle over a fire. The method of cooking is usually boiling or simmering; common foods prepared this way would be soups, stews, beans and other foods cooked in water (vegetables for example). Another common food cooked this way at the time of the Lewis and Clark Expedition was meat. Foods could also be cooked in oil rather than water using this method; today we would call this deep frying.

The second kind of fire irons would be two uprights and a crossbar. One upright is driven into the ground on each side of the fire then the crossbar would be suspended from them. The crossbar, or spit, was run through the food to be cooked. Normally meats were roasted in this manner. A long handled fork was stuck into the roast then the end of the handle laid into a cradle on the end of a third upright so the roast could be periodically turned for even cooking. The crossbar could also be used to suspend kettles and pots over a fire if the meal being prepared requires several pots.

These cooking techniques were very commonly used throughout the country whenever cooking was done over a campfire. The material the "fire irons" were made from depended upon what was available. Although fire irons implies iron rods, it was very common to use wood instead for the tripod and, although somewhat tricky, the uprights and crossbar or spit.

It's almost a given that fire irons were regularly used by the Corps to suspend their kettles and for roasting their meat. If they used iron ones they would have to take them from Camp Dubois with them, but if they strictly used wooden ones they would have cut what was needed along the way.

Since construction of Camp Dubois was already completed it seems quite doubtful that the blacksmiths were making basic construction items such as door hinges for the huts. Andirons and cooking cranes, if they had them, should have already been in place for cooking in the fireplaces. They had not been in camp long enough for any of the iron equipment to wear out and need replacing, other than possibly a small amount of breakage. It seems more likely that the blacksmiths were given a list of items to manufacture in preparation for the trip upriver. Since Clark's journal entries for the time at Camp Dubois were so sketchy the records do not exist, or possibly because they haven't been found, that say what those items were, we are left perplexed.

Although it is impossible to know for sure what those items actually were an educated guess would be metal fire irons (at least the crossbar, or spit, and "S" hooks to hang kettles from the spit) and long handled cooking utensils (spoons, forks, spatulas, ladles) could have been included in the list.

Two ways to use the flour rations the Corps had with them.

Left: dumplings are cooked in a dutch oven over a stew.

Bottom: Bread is made into patties that will be dropped into a kettle of boiling oil along with other meats such as fish, fowl, deer or buffalo

Chapter 16
Myths Exposed

To start analyzing the information we have assembled we will first try to dispel some myths that confuse the issues at hand. The first myth concerning Indians is an overly simplistic explanation of where the Corps obtained much of their food. As is shown it also is in discord with known facts as documented in the Expedition's journals. The fable of volumes of meat consumed certainly grabs attention, but quickly falls apart under objective scrutinization.

Both of the above myths are frequently repeated by speakers and authors, but a third myth was uncovered during research. It apparently is not very widespread. The third myth is unique in that it can be traced to its origin. It is also disquieting because disproving it throws into question much more information from that same source.

Indians Taught Expedition

As the role played by the Indian tribes along the Expedition's route becomes more widely acknowledged myths grow. Documentation of what these people did is often incomplete or only exists in the oral history and traditions of the Indian tribes who live in the area the Expedition traveled through. Consequently it is frequently misunderstood. Some people tend to not want to give full credit while others over-credit. Some common statements made by speakers and writers include:
- Sacajawea provided the Corps of Discovery with many roots and berries during the trip.
- The Indians/Sacajawea showed the Corps how to cook many of the foods they got from the land and the Indians they traded with.
- The Indians/Sacajawea showed the Corps what plants were edible.

A good starting point for understanding why these statements are not true is to read a modest size book, "The Hunting Pioneers" written by Robert Holden. **(1)** The author discusses the first settlers in the Eastern Woodlands as the United States grew westward from the Atlantic seacoast to the Mississippi River. During this period starting in 1720 and lasting until 1840 one of the most well known of the hunting pioneers was Daniel Boone.

These hunting pioneers are frequently referred to a "backwoodsmen" or "white Indians." They lived in rough log cabins with dirt floors and split log roofs. Furniture was almost nonexistent, but what they had was crafted from rough logs. Buffalo robes and black bear hides served as a bed. These people placed a high premium on courage, hardiness, individualism, and personal freedom, and dreaded anything that appeared to be a constraint. Essentials for living were a rifle, ax and knife.

They chose to live in the wilderness for the solitude, so when the neighbors got too close (too close was when you could see the smoke from your neighbor's chimney) they pulled up stakes and moved farther into the wilderness and away from people. This pattern repeated itself every few years. Because life for these people was the most basic of subsistence living, much of their time was spent hunting, primarily deer and black bear. The hunter and his family simplified their needs to the barest essentials. With no desire to own land, they simply occupied a spot in the wilderness. If they couldn't make something they did without. If the hunt was unsuccessful they went hungry.

When one of these hunters went on an extended hunting trip, which they often did, he took some salt and corn meal as his only food; to eat anything else he had to shoot wild game or find edible wild plants. Like the wild animals his life closely mirrored the hunter was indifferent to fatigue and hardship. When trailing his quarry he would track it several days if necessary until he caught it. When nighttime overtook him, he simply lay down and slept. Food was a matter of eating when it was available and going hungry when nothing could be found. Food could be many different things, but usually it included meat from game the hunter shot.

By the time of the Lewis and Clark Expedition the Kentucky recruits and many of the frontier military men had lived all their lives in this environment and knew nothing else. This was the heritage President Jefferson referred to when he set the requirement that the men of the expedition must be good hunters and accustomed to life on the frontier.

As we can see from the brief description above, the heritage of many of the men of the Corps of Discovery was a life much like that of the Indians. The necessary skills and knowledge required to live a subsistence life in the wilds had been learned at an early age. These same skills would be employed during the expedition to the Pacific. These men did not need to be taught how to find edible plants, although the Captains did forbid the Corps from digging camas roots because they were new. The species that were edible and the ones that were highly poisonous were very hard to distinguish. This had little impact on their food supply since they were traveling and didn't have time to prepare the roots so they would keep. The Corps traded for the camas roots they ate and Lewis made a detailed description of how the Nez Perces prepared them. But the Expedition never used Lewis' recipe. The same can be said for Lewis' description of how the Nez Perces pit cooked bear meat. Pit cooking is a very slow, time-consuming method of cooking; the Expedition did not have time to spend pit cooking.

When the Expedition was at Fort Clatsop Sacajawea did extract grease from some elk bones in the Shoshone manner. Clark didn't say if this was new to the Expedition, but he left that impression.

Several times during the journey Sacajawea gathered edible plants, but again the journals leave the impression they were not new plants to the men. She also only collected small quantities, enough for a single meal. One significant exception was while the Expedition was camped with the Nez Perces on the Clearwater River in May of 1806. Clark noted on the 16th that Sacajawea gathered a quantity of "fenel" roots and again on the 18th she gathered more to dry for the trip across the Rocky Mountains. What she really gathered was yampah or Indian carrots. Yampah grows wild in many areas of the United States and today is officially classified as a weed in such diverse areas as West Virginia, Texas, Kansas, South Dakota and Idaho. It is highly probable that the Expedition knew the plant before Sacajawea gathered it in May of 1806, even though Clark misidentified it. Yampah looks quite similar to fennel, so Clark's error probably was lack of close examination rather than lack of knowledge.

The Indians' contributions to the Expedition's food was not teaching them new plants or cooking techniques. Sacajawea also did not teach the Expedition new foods or techniques. She did provide a certain amount of edible plants and she did show them some different techniques, but these men knew the plants and a workable method of gathering and preparing them for food.

Instead it was in the role of providers of food. When the Corps of Discovery, near-starvation, stumbled out of the snow-covered Rockies onto the Weippe Prairie of Idaho they did not want the Nez Perces to teach them how to harvest and prepare camas roots. They wanted them to sell some food. If those Nez Perces had not been friendly or had not provided the Expedition food, the Corps of Discovery would have perished on that spot.

When Jefferson instructed the Corps of Discovery to live off the land he made the natives of the area they were to pass through on their trip to the ocean and back an integral part of the Expedition's success. Without the Indians' friendliness and willingness to trade with the Expedition, the Expedition would surely fail.

Nine Pounds of Meat

Several speakers and writers have attempted to calculate an average daily consumption of meat during the journey for each member of the Corps of Discovery. **(2)** Why such an average would be developed and what purpose it would serve are very curious questions to me.

The attempts at developing an average meat consumption start with trying to determine how much meat was eaten then divide that total by the number of people on the Expedition multiplied by the number of days the journey lasted. This is a simple mathematical calculation, but the actual numbers prove to be more elusive. The only hard number is the length of the journey. The size of the Expedition is questionable and the amount of meat is even more elusive. To get a number for the amount of meat eaten several estimates or approximations are required. Once the number of animals used for food is arrived at, which only can be done by arbitrarily deciding what number to assign to terms like some, several, or a few, then an average weight per animal is given.

Another distinction that must be decided is how much meat the wolves, grizzlies, coyotes and other scavengers got that the Expedition had killed. To better understand this read Lewis and Clark's journal entries for December through February of 1804-5. For example, on December 7th hunters shot eleven buffalo, but the wolves got six of them before the hunters reach them; the Expedition got five of the buffalo. Other entries say when hunters made their kills they took all they could carry and the wolves got the rest before the hunters could return to get what they had left behind.

There are additional problems with trying to determine the amount of meat eaten during the winter at Mandan. These meat-bearing animals vary greatly in weight between seasons. During the winter they are much leaner than during the late summer and early fall. Clark notes on February 10th they shot one elk and six deer and collected all the meat that was fit for use. That winter a common passage in the journal entries right after what animals were shot is "too meager for use." Compare the amount of meat on the two deer Shields shot on February 5th that Clark termed very meager to the one shot in late July the year before that had a layer of fat on it an inch thick.

Still another problem at Mandan was the continual visits between the Expedition members and Indians. Most of the time visitors, Indian and explorers, gave their hosts presents of meat. The amount varied from a modest size to a sleigh load. How is this amount factor into the average?

Historians can't seem to reach a definitive number as far as the size of the group that traveled from Camp Dubois to the Mandan Villages in 1804; the number varies from 45 to 48. Taken over the six months trip, this could be as much as 3,000 pounds difference in meat required.

We have a similar problem with actually knowing how many people were being fed on much of the trip. When hunters were out, particularly on extended trips, does the animal kill include what they ate? Many meals were shared with groups of Indians along the route and many visitors at Fort Clatsop and at the long camp in Idaho waiting for the snow to melt so they could cross the Rockies. Who supplied the meat? The journals do not always give the numbers necessary for meaningful averaging.

A final area that is suspect is to give an average weight for an animal type. The Expedition shot three different species of deer and both sexes of each. What is the weight difference and how can it be determined which species was shot if the journals don't say but the Expedition was in an area where two species live? If it is recorded they shot 36 deer how many were male and how many female?

Since we do not really know how many animals were eaten and even how many people were being fed from what the hunters got, how can we possibly derive a statistically meaningful average daily individual consumption?

An average consumption tells us nothing about their actual diet. In fact it hides the ever-changing nature of subsistence living as the Corps of Discovery traveled across the country. Factors contributing to diet, such as time of year, population density, the Expedition's relationship with inhabitants of an area, and relationships of the groups of native inhabitants all tend to be buried by simply repeating an unverifiable myth that each member of the Expedition ate an average of nine pounds of meat per day. It belies the truths of times of plenty and times of near starvation and the fact that food and how the Expedition secured it was a major factor in the Expedition's successful transcontinental crossing.

The Paul Revere Legend

Brief histories of dutch ovens can be found across the internet on sites owned by dutch oven cooks, backcountry guides and college professors who teach classes about dutch oven cooking. **(3)** Many of these sites include the tale that Paul Revere is credited in the mid 1790s with redesigning cast iron pots by putting three legs, a bail and a tight-fitting lid with a lip to keep the ashes from falling into the food when the lid is removed. In his book "Dutch Ovens Chronicled" John Ragsdale flatly states that Paul Revere invented the dutch oven as we know it today (when he made the design modifications listed above). He further says the only visible difference in these turn of the century dutch ovens from the ones we have today is the quality of the metal, which now is much finer. **(4)** But none of these references give any sources to validate their claim. We can only speculate where this story came from.

Revere, an extremely enterprising and resourceful businessman, was operating an iron foundry in Boston in November of 1788 as well as his hardware store. As Paul Revere expanded beyond being a craftsman into the business world in 1787 he opened an iron furnace. In the 1790s he was producing cast iron stoves as those modern advances started replacing fireplaces for heating homes in Boston. **(5)** There were discussions with the Governor of Massachusetts about Revere getting a monopoly on iron production there. Jayne Triber records that in 1786 Revere received a fifteen-year exclusive privilege of manufacturing iron by "the new invented steam engine." **(6)** By 1801 he had further expanded his interests to include a copper rolling mill and became the first person in the U.S. to successfully produce cold rolled sheet copper.

Revere's legend seems to be based on the general history of how the iron industry was developed in America. When iron deposits were first discovered in the early 1600s they were found in many locations but not in large quantities. **(7)** Nevertheless iron furnaces and forges were built near many of the deposit locations. Although most of them failed, they served the very important niche of developing the technology and training a skilled labor force that would be used successfully later when all the commercial factors were right. By the early 1700s economic quantities were discovered in New Jersey and Connecticut and by mid-century eastern Pennsylvania had become the center of iron activity as commercially economic quantities were found in there that were also near populations. This, along with the business decisions to develop the domestic market instead of shipping it all to England was the start of a thriving iron industry in the U.S. The industry was composed of small furnaces and forges in many localities each filling the needs of the local population. **(8)** The local forges processed the iron into useable products, making all kinds of tools, pots, kettles, nails, etc. Since most of the local forges were involved in making consumer products, the conclusion that Revere's local forge was similarly involved is logical.

However the research never provided any source documents that could substantiate the Claims of Ragsdale and others. I failed to find anything to directly tie Paul Revere to dutch ovens. If he had indeed made such major design changes he surely would have patented them; a search of the U.S. Patent Office turned up nothing other than the fact that much of the early patents had been destroyed or were otherwise missing.

I finally contacted the Paul Revere Memorial Association in Boston. A book which was the exhibition catalog for a major exhibit on Revere's life and work offered few insights into the question other than confirming that he had indeed operated an iron foundry in Boston from 1788 until shortly after the turn of the century. **(9)** The author also confirmed the connection between Paul Revere and Revere copper-bottomed cookware that was produced until the 1950s. Revere Copper Works began manufacturing its well-known copper bottoms for culinary utensils in 1807. **(10)**

Correspondence with the Memorial Association bore much greater fruits. Patrick Leehey, the Association's research director, stated he had never found any connection between Paul Revere and dutch ovens. He did some checking of the extensive collections of Revere's business papers and found no mention of dutch ovens on his hardware store inventory. Although no complete catalog of all the iron products Revere made at his foundry has been compiled from the very extensive business records in their collection the only cast iron ware he found, other than a few miscellaneous objects, was frying pans, but he did find considerable amount of brass ware. **(11)**

An article in the Revere House Gazette and portions of the master's thesis it was taken from shed better light on the iron foundry's scope of operation. **(12)** Since iron ore processing at that time was very seasonal Revere and most other furnaces had trouble getting pig iron during the winter when little if any smelting was done. Consequently everyone relied on getting scrap iron to re-process. Revere produced few small consumer products, concentrating instead on larger goods such as stoves, iron parts of bridges, and other industrial iron goods. As the 1790s drew to a close Revere became what we would call today a defense contractor and supplied cannon and other ordnance and shipbuilding materials to the state of Massachusetts and the federal government. When a hurricane severely damaged his foundry he never rebuilt it. Instead he concentrated on the copper foundry he had built in 1800.

Paul Revere was an industrialist and entrepreneur and apparently used the iron foundry business as a means to what became a much more lucrative copper business. Iron furnaces were widespread and competitive so Revere moved to the copper business where he saw a great need and little if any competition.

Although there remains two small holes that could eventually produce documents to the contrary I am perfectly satisfied that Paul Revere had no connection with dutch ovens. The story of his inventing them is simply a tale.

According to one researcher a modern day dutch oven as shown here was very similar to those used in 1800.

Chapter 17
Field Tests

When researching a subject with as small an amount of documented facts and so great an amount of unknown as the Expedition's food the primary sources are quickly exhausted. Secondary sources and general stories become important avenues to explore, but the pitfalls become greater than is normally the case with secondary sources simply because of the lack of facts to use in determining fact from fiction and theory. It becomes very easy to be lulled into believing a story that is heard over and over again, or to find yourself agreeing with what appears to be sound logic that is being passed off as fact. Research for scientific studies requires time in the lab testing before theories can be finalized. We took this same approach for our research into the Expedition's food and cooking.

Our first field test was simply to cut yucca root into thin slices and drop it into hot oil. After about a minute they popped to the surface nicely cooked to a golden brown. With the addition of a little salt they become a tasty snack. A simple test but it helped prove an important theory.

In the chapter titled "Myths Exposed" the statement is made that the various Indians/Sacajawea did not teach the Expedition members what edible plants to gather or how to prepare them because they already knew what plants they could eat. As they passed through the prairies they collected those plants they knew and ate them. The journals make no mention of several other edible plants we know today they passed by. One of those plants is yucca. Yucca was mentioned in the journals but as a source for soap. Our tests show the root is edible when cooked in oil. Although they had the means of cooking yucca, they didn't do so since they apparently did not know it was edible.

Several other small tests such as this were done to help determine small specific issues or to help us decide what choice to test further. One such test was to see how long it took for oil at cooking temperature to cool down and harden. Another test showed the handle of a kettle of hot oil does not get too hot to hold with a bare hand while the handle of a kettle of boiling water does. These tests support boiling with oil rather than water when time was most critical.

Some major tests that produced significant results are described in the following paragraphs.

Field Test: 6/28/04

The end of June each year the Lewis and Clark community in Great Falls puts on a festival showcasing various local Lewis and Clark significant sites, demonstrations of period skills and lectures on aspects of the Expedition. The Honor Guard sets up an encampment and for three days lives much like the Corps of Discovery did when they were in this area in 1805. The Honor Guard has used this festival that has been held annually since 1989 to not only entertain and educate the public who visit the camp, but to validate what they have learned through research. It is a good time to field test new theories and techniques.

During the 2004 Festival the authors field tested cooking times required for several different methods. The theory we developed was that the Expedition's travel pattern was to go as far as possible each day for a number of days then stop for a few days to rest, repair equipment, restock food supplies, etc. They would then resume their push upriver. During the days of hard travel they were trying to go as far as possible each day. Therefore the methods of food preparation were whatever was the quickest. We had learned for our research that boiling was the common preparation method and the only cooking equipment Lewis purchased that is a matter of record was some brass kettles. But most foods boiled in water are tasteless and take considerable time to cook. Boiling in water also limits what foods can be eaten. We learned through our research the Expedition used a great deal of grease and lard. If boiling in water was the primary cooking technique how cornmeal and flour were used in the quantities the Expedition used remained a mystery. We determined to test two boiling methods; use of water and use of oil.

During the Festival we built two small, comparable sized fires and hung a brass kettle over each; one kettle was filled 1/3 full of water while the other was filled 1/3 full of lard. The testing began. The time required to bring the water to cooking temperature (200-212 degrees) was 40 minutes while the oil got to cooking temperature (325-350 degrees) in 27 minutes. Cooking times for these foods was recorded: buffalo meat strips, fresh or dried, 4-6 minutes; trout, 12" whole, 10-12 minutes; fry bread, 2-3 minutes; squash 20 minutes

Each boiling method has its own limitations. Using water allowed us to boil squash that we could not boil in oil, while using oil allowed us to cook fry bread that could not be cooked in water. We also found that our oil could be used for 14 hours before it burned and had to be replaced. From this information we calculated that each mess would use about 1½ gallons of grease per week when the Expedition was pushing upriver. That gives a reasonable explanation of how the Corps of Discovery use so much grease. We showed in our tests that the foods that could be boiled in oil were closer to the foods named in the journals that the Expedition was eating while on the march upriver than the ones boiled in water. When the Expedition stopped for a few days preparation time was not as critical for determining what foods to eat. So during these rest days cooking techniques could be used to vary the Expedition's diet.

Field Test 10/3/04

Eight members of the Lewis and Clark Honor Guard went to a location on the Missouri River near the Upper Portage Camp for a photo/filming session. One cook was part of the group since the people conducting the session wanted photos of the cook fire and the cook at work. Upon arrival it was quickly realized the group was the same size as a squad or mess of the Corps of Discovery.

The cook, John Toenyes, decided this was a perfect situation to test the time required to prepare a meal for the mess. He selected one man to cut firewood and keep the fire stoked while he prepared the food and cooked (Lewis' order appointing cooks for the messes said they did not have to cut firewood). From the time he started the fire until the food to feed eight men was done it took one hour and ten minutes.

While John was cooking the other six set about putting up a tent and four leantos and otherwise duplicating what the Corps of

Discovery would have done to establish a camp. When the Sergt in charge strode into the cook area to inquire about the possibility of something to feed his hungry men who had just completed setting up the camp, the cook said nothing but handed him a plate piled high with bread, buffalo meat and fish that had just come out of the kettle hanging over the fire.

This test showed us that in the length of time it took to set up camp the cooks could have supper ready. When the Expedition was moving upriver and stopped to eat supper they always had about a hour of work to do which is the same amount of time the cooks needed to get the meal ready.

Field Tests June/July 2005

Extensive testing during June of '05 almost reached a "dress rehearsal" level. The Lewis and Clark Honor Guard re-enacted two Corps of Discovery camps as part of the Lewis and Clark Bicentennial signature event "Exploring the Big Sky." These camps were done in first person, which is the closest way possible for duplicating a piece of history. The authors took on the characters of two of the Superintendents of Provisions (cooks) and for nine days in camp at the mouth of the Marias River then for eleven days in camp at the Upper Portage Camp fed the Corps of Discovery re-enactors using foods, equipment and methods to duplicate as closely as we could what the cooks two hundred years earlier had done in those two camps. We produced some intriguing results that solidified some of our opinions, changed some others, and showed some glaring differences on how we view food today compared to how it was looked upon two centuries ago.

Our plan for cooking was to test everything we had found through our research and the logical conclusions we had made. We would start with a short camp setup (see chapter 20) then go to a long camp scenario. We set up with one cooking fire using a wooden tripod and cooked several meals in a large kettle of oil. We used hog lard since it was most readily available and the Expedition did take 600 pounds of hog lard with them. Other rendered fat oils would have acted similarly. Our meals consisted of buffalo meat and corn bread cooked in the kettle of oil. We varied the meals somewhat by cooking several birds (pheasant and turkey) and fish. We further experimented by cooking strips of ham that had been cured in brine the same way it was done in 1800.

These first tests showed us that we only needed one cook fire since we had only about a dozen people in camp, the same number of people as reflected in the journals. Two or three cooks working together could have easily fed the full Corps, 33 people. (Later at the Upper Portage Camp we did exactly that feeding 30 people in 33 minutes. The meal was the basic Lewis and Clark meal of buffalo, bread, several fish and several pheasants).

We were surprised to learn that the wooden tripod lasted the entire nine days we were in the Marias camp. We had expected we would need to replace it several times because of the constant exposure to the heat from the fire. We concluded that by using oil to cook in we used a smaller fire and the oil absorbed the heat that would have otherwise eventually caught the top of the tripod on fire.

Later in the week we made a pot of ham and beans using a dutch oven set on the ground close to the fire. The cast iron worked like we thought and maintained a good even heat throughout the kettle even though we had a small fire. Further experimentations include boiling squash in water, cooking cornbread in a dutch oven, and cooking stew and dumplings in a dutch oven. We also gathered some wild mint and made tea that was a hit with most of the Corps.

We found that when we varied our meals so that a single kettle of oil would not cook everything we needed a larger fire either to get enough coals for the dutch ovens or to heat the kettles set on the ground near the fire. With a larger fire our wooden tripod suffered leading us to believe that the Expedition would have used a second fire or metal fire irons.

We did not spit roast any meat during either of our camps since we had done that many times in past years. If we had decided to roast some meat we would have needed to increase our fire (and probably use iron uprights and spit). This would have prevented further testing we were doing using a single, small fire.

Field testing our ideas was an essential part of our research.

Left: testing boiling strips of meat in hot oil using a copper kettle

Bottom: A whole trout is removed from hot oil (it tasted mighty good)

Chapter 18
Weighing the Evidence

So far we have seen a short list of cooking equipment we know the Corps of Discovery had because purchase receipts still exist for them. We have discussed foods we know they took with them, others that they gathered from the land and those they traded for. These are reasonably well documented by the journal keepers. The journal keepers told how some of the foods were prepared.

Next we discussed some possible additional cooking equipment and cooking methods that seem likely by looking carefully at what is written in the journals. When we add those results with the fact that Lewis' initial purchases were for a unit of a dozen while the unit that set out from Camp Dubois in the spring of 1804 was nearly four dozen we become fairly sure some additional cooking equipment had to be obtained.

We have exposed three myths that cloud or distort the issues of food and cooking equipment then reviewed our field tests that were done to help us better understand the research information we gathered and to possibly fill in the gaps in the journal record. By re-creating situations during the journey we tried to re-create the commonplace activities the journal keepers didn't bother to write about.

After making many small tests then doing more extensive testing, especially during the two camps in June of '05 we are quite certain of our conclusions regarding how the Corps of Discovery went about their cooking chores. We are convinced we came about as close as possible to duplicating their methods, food and equipment. Our field tests produced another result we were not expecting that completely surprised us.

After the signature event was concluded the Honor Guard was doing a critique of our month's activities. One of the men brought up the topic of food by stating he hoped that in the future we could vary our meals to be somewhat less authentic. He suggesting including more salads, vegetables and desserts along with a little less boiled meat. He concluded by remarking that it was very intriguing to participate in the historically accurate way, but a big piece of pie sure would taste good.

A few days after the critique session was concluded the impact of these comments more fully hit home. The authors discussed how food was regarded by the hunting pioneers of the late 1700s who's only care was to get something to eat; they were very indifferent about what it was and cared little if it was the same diet for weeks on end. Today we have come to demand a great variety in our diets. The change has become so significant that we wondered if the Corps of Discovery would have shared our demands for variety today would that have caused the Expedition to fail.

This concept becomes important when analyzing the Expedition's cooking habits. We realize we can use our attitudes about food to determine acceptability. They may not have used taste as the primary consideration for how food was prepared or what foods were cooked. Other factors became more important; what the land was producing at the time they were there and the time required to prepare it, along with the easiest way to prepare the food, could more easily replace taste or variety if these two attitudes are not important.

The final item we discussed was a "military unit theory" commonly used to fill in the missing pieces and to answer the questions of additional equipment and cooking methods. Our task now is to carefully examine that theory to see if it actually does the job of giving the answers to our basic questions of food preparation methods and equipment used.

We determined the Corps of Discovery was a military unit from its inception by President Jefferson until its return to St. Louis in 1806 and the men were mustered out. But can we truthfully say that they always acted like the stereotype military unit?

The Corps of Discovery appears to be more like a modern reserve unit; part time civilian and part time military, but carrying certain civilian attitudes. Or more like a Special Ops unit than an 1803 Infantry unit. The unit's mission was to go do a job and not return until it was completed, or the unit was defeated. By contrast, normal army operations of that era were carried out during the "season" with the unit returning to winter quarters until the next "season", but this new unit would be gone more than a year (almost 2 ½ as it turned out). **(1)**

The Corps of Discovery had co-commanders which was unheard of in the normal military unit. Some historians speculate that the War Department would only give Clark a Lieutenant commission because they reasoned that if he were commissioned a Captain it would create problems of who was in command. Although Clark was technically a Lieutenant and Lewis was a Captain, in the field they were equal in command. Both Captains and the men referred to Clark as "Captain."

During the winter of 1803-1804 while the men were being assembled at Camp Dubois a large measure of their food came from the hunters' rifles as they lived off the land. This method of obtaining food continued throughout the life of the Corps of Discovery. Regular military units in 1803 simply did not do this.

The records are somewhat deficient, but from what still exist we can deduce that although the Corps of Discovery was to be gone for an extended period of time they received only the normal issue of clothing with the majority being what the men that were already in the army brought with them from their old units. This fact meant that later, during the journey west, when the uniforms needed replacing the men had to make their own clothes from whatever resources were available; animal skins became the choice. As early as the winter of 1804-1805 at Fort Mandan the journals record the men were making clothing items from animal skins, starting with moccasins to replace regular shoes. Probably by the fall of 1805, and certainly by that winter at Fort Clatsop, this infantry unit was clad almost exclusively in buckskins instead of military issue uniforms, which had worn out or rotted so badly they fell apart. We can only speculate on the impact the lack of uniforms had on the "military attitude" of the Corps of Discovery. We find no degenerating of the discipline of the team that had been so carefully molded by Clark at Camp Dubois. Surely the command structure remained intact, but what impact was made on how the men thought through problems and fulfilled their basic needs of food?

The two Captains did not maintain a steadfast adherence to military procedures. When they drafted their plan for how they were going to establish military courts during the journey they broke with War Department requirements of and put enlisted men on the Board. **(2)** The War Department specified officers only. Lewis and Clark held several court martials using enlisted board members and punishments were ordered and carried out. The Captains did not always follow War Department requirements for punishments required for certain offenses. Instead they determined what was appropriate on their own and even decided if the punishment ordered would be suspended. A good argument can be made that the court martials were illegal.

From these discussions we can see that both Captains freely adapted their actions to meet their needs to get the mission satisfactorily completed. The degree they acted as a normal military unit depended upon resources available and the needs of the mission. A number of writers and historians have suggested that the success of the expedition was in no small part due to the Captains' flexibility and willingness to make changes as needed. Most protocols and customs were left by the wayside if they hindered the effort to get the mission accomplished.

In view of the above it appears logical to conclude that although the Corps of Discovery was a military unit that fact alone did not dictate how the men or their leaders conducted their affairs. One of the areas they most commonly broke with military tradition was in obtaining food. The variation from tradition is to be expected considering the units mission; there simply was no alternative but to live off the land. It should come then as no surprise that their food preparation techniques and equipment would not be in strict compliance with military tradition. If the type and quantity of food can't be predicted it follows that the preparation can't be predicted either. At least some of the foods surely would be new to the men so how to prepare it would be new also. That could easily mean different cooking equipment than standard issue would be needed. It is a matter of record from archaeological excavations at frontier fort sites that military units frequently violated standing orders on food preparation, so given Lewis' order of July 8, 1804 that gave the cooks discretion to use their own judgment on preparing the food, it should be no surprise the standing army orders were openly violated.

Analysis of some of the other arguments presented in the Military Unit theory shows that theory to be overly simple. One of the arguments against any cast iron dutch ovens or iron tripods/fire irons is the excess weight they would add. However From the time Lewis left Pittsburgh until the Expedition left Camp Dubois in May of 1804 the number of boats was adjusted to meet the needs of the cargo. He started with a keelboat and a canoe, but added another canoe and a pirogue as cargo and water levels dictated.

During the winter spent in Camp Dubois Clark spent much time developing the right number of people for the group that would set out up the Missouri in the spring. By reading Clark's journal we see he tried as many as three pirogues and the keelboat before finally settling on two pirogues and the keelboat. From these notes it looks like the primary consideration was the number of men, supplies and equipment needed to do the job not the cargo weight. Clark would simply add enough boats to haul the resulting cargo. Since their trip was to be almost entirely by water the weight could be handled by the number of boats they took. At that point in their planning they thought they would only have a very short, half-day, overland portage from the waters of the Missouri to the waters of the Columbia. Normal army travel was over land with the soldiers carrying their equipment so weight was critical. This is another example of how the Corps of Discovery differed from a stereotype infantry unit.

Another part of the Military Theory says that any additional items the Corps of Discovery obtained, if they got any more at all, came from the public stores at forts along the way. There are no records of purchases because any additional items from the public stores were not purchases but transfers of items from one military unit to another so no records were made. This response immediately becomes suspect since the Quartermaster books for 1780-1781 at Fort Jefferson show many transfers of equipment to other units and unit commanders for redistribution to their men (3) The quartermaster had to account for their inventory so they maintained a written record. A better answer appears to be that if equipment was obtained at the frontier forts along the way the records either have not been found or have been destroyed.

If Lewis couldn't get the very modest requirements of his initial list of required cooking equipment from public stores in Philadelphia and had to turn to private merchants, would he be able to get them in the small frontier forts he visited during his trip along the Ohio and Mississippi Rivers? The two main forts Lewis visited were Kaskaskia and Massac. The commanders at these forts had been directed by the War Department to help Lewis with his manpower needs, but nothing had been said about equipment. **(4)** Of course there would be no need to talk about equipment since Lewis had already made his purchases at Philadelphia. Each of these forts had about 100 officers and men assigned, so the quartermaster's supply of equipment for cooking could not have been very large. **(5)** Even considering he did obtain cooking equipment from these forts how could he have known how much more to get since the final size of the Corps of Discovery wasn't determined until late March the following year. When Lewis was at these forts his mindset was a party of a dozen men and he had already obtained the necessary equipment. Since the merchants in St. Louis routinely outfitted trading parties going up the Missouri, and since Lewis had an open-ended letter of credit from the federal government, it seems more probable that Lewis purchased any additional equipment from them, rather than obtaining it from the frontier forts (those forts probably had precious little, barely enough for themselves).

Another argument against cast iron dutch ovens and iron tripods is simply the expense. But remember that although Congress had appropriated a modest $2,500 for the Expedition, the President had given Lewis a letter of credit with no dollar limits. While Lewis was making his purchases in Philadelphia he was undoubtedly working within that budgetary limitation. However by the time he would have been making additional purchases in St. Louis he was equipping a much larger force; he would surely have looked at his budget differently. A few cast iron kettles would hardly have been noticed or questioned by auditors. These items, along with iron cooking spoons, forks and spatulas are regarded as extravagant by the modern militarists. They say brass kettles can be used to fry in and even bake in by fashioning a lid or by inverting another kettle of the same size and placing it on top of the kettle with the food in it. But a distinction must be made between making do for a day or so of primitive camping and a year or two of travel in the wilderness.

For short periods of time or for a small number of occurrences most of us are willing to make do with as little as possible. However the Captains must have had a much different mindset as they prepared for their journey since they knew they would be gone for an extended period. We have already seen that Clark spent considerable time balancing the equipment, men and boats requirements. A better description of their mindset probably would be to purchase or make what was reasonable to include in the equipment for their trip. This was not a "make do" trip; they were not planning a test of their woodsman or survival skills. They were planning to make an extended trip exploring the western lands of the Missouri River and included the things they reasonably needed to get the job done. With their camp located in close proximity to St. Louis and merchants who were familiar with equipping groups going into the wilderness, along with an open letter of credit from the President, the Captains had access to what they reasonably would use and the means to pay for it. A minimalist, survival mode was not necessary.

One highly regarded "dutch oven cook" expressed serious doubts about frying or baking in brass kettles. **(6)** She says brass kettles require a much hotter fire than cast iron ones. Not only would the high heat required be much more likely to burn the food being baked, it could easily start melting the solder on the brass kettle and make it leak. A hotter fire would require more wood, but the planners knew they were going through an expansive plain where wood supplies may well become very limited (this is exactly what they found in eastern Washington). Heat is evenly dissipated throughout cast iron while copper or brass develops hot and cold spots that would add further to the problem of food burning. Cast iron tends to hold its heat longer than copper. Heating cast iron is a "cumulative" process; the characteristics of it is that it holds onto its heat while heating copper is an "additive" process; copper tries to get rid of its heat.

A basic argument of the military unit theory is that the army regulations required foods be boiled. Brass or tin kettles were the only "cooking pots" issued since they meet the army needs. Frying pans (spiders), roasting spits and racks, or cast iron dutch ovens were not issued. With this knowledge the militarist concludes the Corps of Discovery did not have any of these items. In fact some say the only cooking equipment the Corps of Discovery had was that each mess had three brass kettles. Acknowledging the Corps did on occasion roast meat, they say spits were made from green willows gathered around the camp; this was commonly done by civilians on the frontier. They also say that the few times the Expedition fried their food it was done in a brass kettle. This argument ignores the fact that in Lewis' order of July 8, 1804 that established a Superintendent of Provisions for each mess also stated that these cooks were free to exercise their own judgment on how to prepare the food. Since most of the time the specific mode of cooking is not revealed in the journals, it becomes an argument of frequency. On one hand we have learned by experimentation that boiling in oil is the fastest method the cooks had at their disposal. But balancing that fact is the knowledge that other preparation methods usually give better taste to the food, particularly compared to meats boiled in water. The military unit theory relegates any methods of preparation other than boiling to a "seldom used" category, which, in turn, supports the contention that "make do" equipment would have been used. That theory also ignores the question of how the flour and cornmeal was used.

In response to the journal reference to caching a dutch oven(s) at the Marias, the militarist says we simply do not have the information needed to tell what they cached. However some do acknowledge that the Expedition deposited their heavy items, such as a dutch oven, before starting their journey overland across the Rocky Mountains.

There are several holes in this argument. Although it is true they cached some of the heavy items, they also cached flour, corn meal and other provisions for their return trip. At that point the Captains still thought they would have a short portage between the waters of the Missouri and the Columbia. They traveled another 50 miles upriver before they discovered a difficult 18 mile portage was required to get around the five falls of the Missouri. But that portage put them back on the navigable waters of the Missouri for an additional 150 miles.

Upriver 50 miles from the Marias the Expedition camped and prepared for the portage around the falls. Recent archaeological excavations at that camp (called Lower Portage Camp) revealed several holes about ½ inch in diameter and 3 to 4 inches deep had been poked in the ground near a fire pit. The mystery of these holes was only solved when a dutch oven was positioned over the holes; the legs of the dutch oven aligned perfectly with the holes.

Cast iron frying pans, with and without legs, were common cooking utensils. We discovered that Paul Revere sold them in his hardware store in Boston. The legs on a frying pan would make holes very similar to those made by a dutch oven. However a dutch oven suspended over the fire suitably duplicates a frying pan so there would be no need to take extra equipment. Since there is no documentation showing frying pans were taken and our research showed no need for them, we seriously doubt frying pans were taken on the journey.

After the portage was completed and the collapsible iron framed boat that was to replace the pirogues failed, the Expedition continued their journey in dugout canoes. Additional items had to be cached at the Upper Portage Camp because everything would not fit in the dugouts. Lewis complained about how much extra the men were collecting along the way and how much trouble he had getting them to leave things behind in the cache, but no mention was made of dutch ovens in the list of items cached there. The utensil that left the holes by the fire at Lower Portage Camp may very well have been taken with them on the journey through the Rocky Mountains.

Since we can't use the concept that the Corps of Discovery was a military unit so they did what a military unit would do to answer all our questions of what, if any, other cooking equipment they took, we need to return to our discussion of several items. We have discovered that Ragsdale's statement that Paul Revere invented the dutch oven as we know it today has been shown to be in all likelihood in error. Although he provided no source documents as proof of his conclusion, we stop just short of saying he was wrong. We could find no sources to prove him right or wrong, but everything we found shows Revere had no connections with manufacturing dutch ovens. Therefore we are satisfied to say Revere had nothing to do with dutch oven design or manufacture.

Knowing this error exists in his book that has become the basis for modern era dutch oven "history" we questioned several other aspects, in particular what they looked like by 1800. (Ragsdale may have concluded that Revere invented the dutch oven from his research and never meant it to be taken as a fact; if so he erred by not clearly marking it as his theory.) Verification of his statements from other sources became important.

Verifying Ragsdale's work was the most difficult research done for this book. He left no substantive record of his own research and most everything I could find referenced his book. Other writers repeated Ragsdale; right or wrong. Consequently I was unable to find independent sources to be able to say "yes" or "no" that he was right. So the confusion continues. What was a "dutch oven" in 1800 and what did one look like?

What was outlined in chapter 15 on dutch ovens makes the best logical sense, including how this heavy iron pot became known as a dutch oven. Ragsdale's work being repeated on the website of current dutch oven manufacturers lends some credibility to his results. Except for the arguments outlined in chapter 16 under "the Paul Revere Legend" the authors accept Ragsdale's work, with reservations.

Chapter 19
Formulating Theories

In previous chapters we have discussed the list of foods the Expedition ate and where they came from. We have also discussed how these foods were prepared, discovering some were roasted, others fried and many were boiled. There were even some that were eaten raw. And we have spent considerable time trying to determine what equipment and utensils the Expedition had with them to use in preparing their food.

The one thing common to most of our discussion has been the fact that so little was written on the subject of food by the journal keepers or others of the Corps of Discovery at the time of the journey. Consequently we have had to go to related documents and analyze them in the light of the methods of operation of the Corps of Discovery to try to find the missing pieces. When searching for additional records we were frequently thwarted because the hoped-for records could not be found. Quartermaster records from Fort Massac or Fort Kaskaskia, for example, which could have shed more light on equipment that Lewis may have obtained or records from merchants in St. Louis to further explain the "sundry items" Clark purchased. Many of the early patents issued by the U. S. Patent Office were destroyed by fire, so we could not determine from that source the validity of the legend that Paul Revere modified the dutch oven to be what it is today. A more circuitous route that involved three books on the life of Paul Revere and an additional one on the history of metals yielded the story.

When sources can't be found there is always the question of whether the researcher didn't look diligently enough or the record simply doesn't exist; whether no written record was ever made or it was made and later destroyed. This is the dilemma we faced.

The bottom line statement that everything has to be documented or it can't be included is simply not realistic. This "one size fits all" declaration may work in a black or white world, but reality contains many shades of gray. We know there are many holes in historic fact because documents have become lost or destroyed (consider how much was lost when the English burned Washington City during the war of 1812) and because people simply didn't write every detail of every aspect of every topic. Journalists throughout history omitted the commonplace, favoring the more uncommon or what they in particular did or thought.

The military unit folks are correct to say the only things that can be included AS FACT are what's documented, but they can't stop at that point; they must continue by allowing for sound theories to be used to complete the picture. However, these theories must be labeled as theories.

In our discussions of other equipment, any theory put forth must be tested by opportunity and knowledge. Questions to be asked include did the item exist at the period of time being considered; was it available to the people being considered; did the people know about it and how to use it; would the people want to have the item; would the item be beneficial to be used by the people. If the answers are yes then the theory is probably valid.

Theories may never become facts if the sources required to prove them can't be found. This doesn't discredit the theory. The test of a valid theory is if it presents the known facts and is logically developed to a reasonable conclusion. And it must meet the test of workability. The facts of most of our history are full of holes that can only be completed by theories.

We found that there were a variety of different ways to do almost every task. Our job was to discover the most likely way the Expedition did things. In the following pages we will present our conclusions based upon our research, field-testing and analysis.

These are the theories we have developed that fill in the holes in historical fact and tell the story of how the Corps of Discovery prepared their foods.

- *Theory 1: The primary factor influencing cooking techniques was the amount of time available.*
- *Theory 2: The Corps of Discovery took cooking equipment in addition to what is documented in receipts from Philadelphia merchants. Additional equipment was purchased from*

merchants in St. Louis including brass kettles and dutch ovens. The Corps had long handled forks, spoons and spatulas they obtained through some combination of purchases from St. Louis merchants and production by the blacksmiths while at Camp Dubois. They also took fire irons, but only the crossbar, or spit, and hooks to suspend kettles from, the blacksmiths made while at Camp Dubois from iron stock purchased in St. Louis. The men who joined the Expedition brought with them personal eating utensils consisting of plates, spoons, forks and cups. No additional equipment was obtained from the frontier forts Lewis or Clark visited on the trip to Camp Dubois.

- Theory 3: The principle cooking technique the Expedition used was boiling although they frequently roasted, fried and baked food particularly during their longer camps and during the winters at Fort Mandan and Fort Clatsop. The food, when boiled, was boiled in oil (lard, bear grease, buffalo or elk tallow) whenever it was available otherwise it was boiled in water.
- Theory 4: The messes all ate and cooked in a similar fashion including the mess formed by the Captains. We do not know who cooked for the Captains mess, but it was probably a combination of Charbonneau and York.
- Theory 5: the messes combined at times to simplify cooking, especially when some of the men were out of camp.

Our Theory of Time

The Lewis and Clark Expedition was a military unit with a specific mission to accomplish that required an indefinite amount of time to cover an unknown area traveling primarily by water with a brief overland portage. To succeed the Expedition had to cross a major mountain range and suffer through at least one winter in the unknown lands. Consequently they had to adjust their travel to be at their winter camping area, Fort Mandan, at a specific time. They also had to be across the Rocky Mountains before winter set in and snow blocked their passage. To make each of these destinations they were required to travel considerable distances with many unknowns, any of which could prevent the travelers from reaching their goal. The best way to help themselves reach their travel goals was to go as far as they could as fast as they could each day. That meant minimizing as many obstacles as they could.

Since they could not possibly carry all the food they would require for the duration of their journey, they were obliged to obtain most of it from the land they passed through. The daily process of securing food slowed the Expedition's progress, but was beyond their control. What they could control was the time required for preparing their daily food needs. The Captains affected this in several ways. They specified that when they were traveling only the evening meal would be cooked; other meals would be leftovers.

Living off the land means a basic subsistence diet. The simpler the diet the quicker and easier it is to prepare the food. For example roasting chunks of meat from a freshly killed buffalo is easier and quicker than carefully aging and marinating choice steaks then grilling the meat the perfect length of time for each diner's taste.

The Captains also established the daily routine of early departure from camp, normally shortly after sun up, and traveling all day until about 5:00 pm before stopping long enough to cook supper. Supper was cooked and eaten then everything repacked and the Expedition would push off again in two or three hours. They would then continue traveling until dark before stopping for the night. This stop was to sleep; no cooking was done. The next morning the Expedition was up at daybreak for a repeat of the day before.

They had to go as far as they could each day because they never knew when something was going to happen to slow them down or stop them completely. During the trip across western North Dakota and eastern Montana in early 1805 winds caused delays many times for part or the entire day. Consequently in everything they did each person had to be very time conscious. Their cooking and eating habits reflected this same consideration. They could vary their diet when they had to stop for a few days, but when traveling it was "make do as best they could."

Our Theory of Additional Equipment

In chapter 4 we looked carefully at several journal passages that gave some hints about cooking equipment. Words like "utensils for cooking" are general, but normally refer to such things as spoons, forks and spatulas used for cooking. These are to be considered separately from personal utensils, which would be the items used for eating with. One of the most convincing references is contained in the April 1, 1804 order that established the messes. It says "the camp kettles and other public utensils for cooking" would be divided among the messes.

There are references to mishaps with the dugouts and the white pirogue where the journals mention that cooking utensils were lost overboard. If these utensils were referring to forks, spoons and spatulas (and we are convinced they do) they must have been made of metal. If they were wooden they would have floated and probably been recovered.

Because of the journal references as indicated in chapter 4 and because of some foods they ate required specific cooking methods we concluded more equipment than what is recorded in purchase documents form the Philadelphia merchants was taken on the journey. This conclusion is supported by the fact that Lewis made his purchases in Philadelphia for a party of a dozen, but almost four times that number actually set out on the upriver trip from St. Louis.

Although Lewis purchased pint tumblers and iron table spoons in Philadelphia the only further mention of either is when Clark issues "tin cups" (presumably some of the tumblers Lewis bought in Philadelphia) to engages at St. Charles. Reflecting that standing military policy was for the men to furnish their own personal utensils and most of the men were currently members of the army when they joined the Corps of Discovery we conclude the men followed military policy and supplied their own. The iron spoons and pint tumblers may have been issued to the nine civilians recruited by Clark. We can find no compelling evidence personal utensils were totally supplied by purchases made in Philadelphia or elsewhere. Each man probably brought his own spoon, plate and cup with him when he transferred from his old unit to the Corps of Discovery. As a result these items were a collection of various types of wooden and metal (tin plated sheet iron would have been the primary metal). Cups were probably plain tinplate with a handle (no tankards) that were common to the period. Lewis' order for tumblers without handles provided for more compact packing during travel but do not set a standard for determining what the men provided for themselves. Forks were iron wire twisted with two tines, although everyone may not have used forks. Spoons were common wood-carved type and may have been more universally used than forks since one item could be used for multiple functions. (You can't eat soup with a fork).

We have concluded that none of the men brought any camp kettles or tin pots with them from their old units. These were unit issue so individuals could not remove them when they changed organizations like they would personal or individual issue items. Individual issue items would be their uniforms and weapons. Lewis purchased the brass kettles to serve as the Corps of Discovery's camp kettles.

Because the size of the Expedition grew from twelve initially planned to approximately 45 and since these were the primary cooking utensil we are convinced additional brass kettles were purchased for the trip. If six were thought to be the required number initially we can't believe that Lewis or Clark would have later decided that number would be sufficient for the much larger crew.

Trying to establish an exact number of additional kettles becomes rather tricky. If the number of kettles on hand is traced through the journals, Lewis' purchase of 14 in Philadelphia and no additional purchases made is correct. However, that number makes no provisions for Corporal Warfinton's return party having any kettles to cook in when they took the keelboat back to St. Louis from Fort Mandan. It further provides for only seven kettles for the crew of 45 from September 1804 until the engages were paid off and left the Expedition in early November. One kettle for every five men may have been satisfactory except all the kettles were not the same size; they varied from one to five gallons.

The next problem created by adhering to the numbers in the journals is at Fort Clatsop. When the saltmakers set up their operation they took all but one small kettle. How did the Expedition boil enough elk meat to feed 28 people in a single small kettle?

We must also consider whether Lewis provided any cooking equipment for the engages who were organized into a single mess during the trip upriver to Mandan. If so, what happened to that equipment when the engages left the Expedition?

Ken Wilk with the Army Corps of Engineers has suggested that each mess had three kettles at the start of the trip. That would be a total of 18 kettles, which is a logical number. This number presumes that the Captains acquired four additional kettles somewhere. It also presumes the seven kettles given as presents were being used until they were given away and afterward each mess was reduced to two kettles.

During our research we could not find adequate evidence to establish a good number for additional kettles. The argument for three per mess is compelling as it provides more reasonable numbers at Fort Clatsop and could fit into the numbers traced through the journals. The best we can conclude is that the Expedition took additional brass kettles, probably at least four more.

We concluded the Corps included in their cooking equipment some dutch ovens, probably several. They cached some at the mouth of the Marias River, but took at least one with them up to the Great Falls. The evidence of the leg holes found during the archaeological excavations at the Lower Portage Camp in compelling. Because Lewis recorded cooking a large suet dumpling for each man while at the Upper Portage Camp we suspect he had at least two dutch ovens there. They probably were cached at the Upper Portage Camp to minimize the cargo of the dugout canoes when the iron boat failed.

We are further convinced the Expedition did not take frying pans or "spiders" on the journey. We can find no evidence to the contrary nor can we rationalize a need for such since a dutch oven can easily serve the same function as a frying pan.

Several journal entries refer to utensils for cooking which we have concluded means long handled metal forks, spoons, spatulas and strainers. Since these were commonly used when cooking over an open fire we believe the Corps included them in their inventory of cooking equipment. The blacksmiths made these utensils at Camp Dubois in February and March of 1803. The blacksmiths also made iron crossbars (or spits) during that time that also were taken on the journey and were used during the long camps and winter camps. Our testing showed us they had no need for iron tripods or uprights for cooking. They used wood for the tripods they regularly used to suspend their cook kettle over the fire. During those times they were traveling in areas of little or no wood they carried the wooden tripods with them. When they roasted meat on a spit they used wood for the uprights and iron spits. Detachment order of July 8, 1804 exempts the "Superintendents of Provisions" from collecting firewood or forked poles for cooking.

Our research has led us to conclude that additional cooking equipment was obtained by Lewis and Clark, primarily because the size of the Expedition grew four fold. The Captains did not determine the exact size of the unit until late in the winter at Camp Dubois. Clark spent considerable time balancing the size of the crew and the amount of food and equipment to take and the number of boats that would be needed for transporting everything. We found that the size of the military forts in the area were all small, hardly larger that the Corps of Discovery. Because of their size they undoubtedly could not provide the additional equipment requirements; especially when it is realized that Lewis could not even get his initial purchase from public stores in the Philadelphia area. Add to that the fact that Lewis had an open letter of credit from President Jefferson that would have made purchases from civilian sources possible. Given the above and the nearness to St. Louis whose merchants were accustomed to outfitting trading parties that were traveling upriver as the Corps of Discovery would be, in the absence of any quartermaster records from the army forts, we concluded that all additional equipment purchases made by the Captains came from the merchants in St. Louis.

Our Theory of Cooking Methods

From reading the journals it becomes evident that the most common cooking method for meat was boiling. The military standing orders specified boiling meat as the method of cooking. Additionally civilians living on the frontier commonly used this method. Although, people of the time did not understand the concept of bacteria, spoilage or high level bacteria counts the military knew that when the meat was boiled before they ate it kept the men in better condition without dysentery and other sickness.

In the twenty-first century we understand bacteria and that we need to get the temperature of our food to a temperature that will kill the bacteria before we eat it. When the meat is boiled it is brought to about 200-212 degrees—near the water's boiling point. Since most bacteria die at about 165 degrees, this was a very good practice. The only problem with boiling meat it is not very palatable. When fresh meat was available other methods (roasting and frying) were probably used just from the fact of a better taste. Grease or oil (bear grease) could have been used to actually fry meat in oil like a deep fryer. Meats cooked in bear grease were tastier and mentioned in the journals. White pudding was boiled then fried in grease. The men cherished the "white pudding" or "boudin blanc" which was a sausage. Cooking it in grease along with the amount of fat included in the sausage is what improved the taste.

Two entries in Lewis' journal illustrate this point. On Christmas of 1805 at Fort Clatsop he lamented that their dinner consisted of boiled poor elk and hoped for much better in time to come. Another entry made June 26, 1805 when the Expedition was portaging around the Great Falls of the Missouri states that he boiled large quantities of excellent buffalo meat for the portage crew. What caused such a difference in attitude towards a meal of boiled meat? It surely couldn't have been elk vs. buffalo since they ate freely of both types throughout their journey.

We know that the Expedition left Camp Dubois with a good supply of lard and grease; they had hardly more than started their trip when they purchased 600 pounds of grease from some traders they met on the river. We are told in many journal entries that the men rendered bear fat as well as buffalo and elk tallow at every opportunity they got. They literally used several thousand pounds of oil. In fact while at the Great Falls they rendered three barrels of buffalo tallow.

By contrast by the time the Corps of Discovery reached Fort Clatsop they were out of oil. During their eleven day struggle through the Rockies they ran out of all their food supplies. The trip down the Columbia afforded scant opportunity to do any hunting so they could not replenish their supply of oil. The difference appears to be the fact the meat cooked at the portage was boiled in oil while the meat cooked at Fort Clatsop was boiled in water.

A test we conducted supports this conclusion. We boiled pieces of meat in water and compared them to meat boiled in oil. The difference was obvious.

Not only did our testing show an obvious difference in taste, but we also discovered meat boiled in oil cooks in almost half the time required for boiling in water. That fact makes sense because water temperature for cooking is approximately 200 degrees while oil's temperature is 325-350 degrees. We found that the time required to get oil to its cooking temperature was less that the time needed to get water boiling. An interesting twist came about during our testing. The handle of the kettle we were using to boil water in got hot as the water reached boiling temperature, but the handle on the kettle we were boiling oil in remained cool to the touch even at nearly double the temperature. This fact vividly illustrates the different natures of the two liquids. The nature of oil is to absorb and retain heat while water dissipates the heat. Consequently heating water can be seen as a replacement process while heating oil is a cumulative process, which explains the times we recorded during our boiling experiment. As a result a larger fire producing greater heat is needed to boil water.

We concluded from our research and tests that boiling with oil rather than water requires a smaller fire, produces better tasting meat and takes less time. An extra benefit is that other foods can be cooked in the oil at the same time the meat is cooking. Boiling in oil is consistent with the references to what foods they were eating. If they were boiling in water the question of how they used their supply of flour and cornmeal is unanswered.

Boiling is the most frequently mentioned cooking method, but the journals specifically say other methods were used; they say they roasted and fried some of their foods. We are told the cous and camas roots were usually eaten like mush. Some of the foods they ate, by their very nature, dictate the cooking methods; suet dumplings and tarts are baked.

Because the Expedition wrote very little about the methods they used to cook their foods the issue of frequency of use of the various methods is important. If the number of times each method was mentioned is used more like a survey and the percentages are applied to the entire trip the frequency of roasting, frying and baking becomes more significant. (To illustrate say the method of cooking is mentioned 100 times and frying is mentioned 10 times, roasting 19, and baking 8. If we conclude that 8% of their meals included baking that becomes a significant amount of baking, rather than just saying they only baked 8 times.) We can conclude they used these other cooking methods frequently enough that they could well have planned in advance to include equipment that would make cooking easier than it would be in a "make do" situation.

Our Theory on Similar Cooking

During our research we found no indications that each hunter was assigned to strictly provide for his own messes. In fact, many of the references only say hunters were sent out so we don't know if all the messes were represented at any given day. Even if they were they would split up so some hunters may not get anything while others would find plenty. Several references leave the impression that the men were given, to the greatest extent possible, comparable amounts and kinds of food, whether it was from the St. Louis rations or from hunting or trading. The orders for May 26, 1804 and July 8, 1804 establishing the messes and superintendents of provisions are good examples. There are also a number of references of one group leaving some of their freshly killed game for food for another part of the Expedition. The various entries that refer to hunting never talk about how the kill was divided. One can only presume that an equitable distribution was made, to include the Captains' mess. During times of plenty everyone seemed to eat sufficiently and distribution was not a problem.

There is some evidence to point to the fact there was not a totally equal division of food. One example probably was when Sacajawea gathered small quantities of various roots or berries. Most of these were given to Clark as samples or in such limited quantities that it stayed in the Captains' mess. The two birthday meals Clark ordered may have been specially cooked for him, while the other members of the Expedition may or may not have eaten similarly. We concluded the birthday meals Clark ate were different from the others because of the extra time and effort that would have been required to cook similar meals for everyone and since there is no mention of anything akin to a "party;" Clark simply had York prepare him his special meal.

When a limited number of certain, smaller animals were included in the day's hunt they probably went to the mess the hunter who shot it belonged. However, if the day's hunt was limited to a few small animals and there was no other food, what there was probably got common distribution. Unless several turkeys were shot, we can assume the Captains' mess got it since Clark talks about shooting and eating turkey several times.

The important issue in this discussion is that although during the 1700s British generals mostly ate very well with little or no regard to how their junior officers and enlisted men fared, this was not the case with the Corps of Discovery. The philosophy stated by the Mandan Chief Big White not only summed up the relationship between the Corps of Discovery and the Mandans, but it addressed the situation within the Corps itself. He said, "If we eat you shall eat, but if we starve you too shall starve."

With much evidence suggesting equitable distribution of foods and no evidence to the contrary we conclude the messes, including the Captains' mess, ate mostly the same kinds and quantities of food and prepared it the same way using the same equipment.

Our Theory on Combined Messes

The Corps of Discovery was a military unit and was organized in the typical military manner relating to how the men were fed; they were grouped into messes of six to eight men. During the Expedition's travel from Camp Dubois to the Mandan Villages where they spent the winter the organization probably was closely adhered to. If on occasion the assigned cook for a mess was unavailable another member of the mess filled in, as provided for in the Captains' organizational orders. Cooks being unavailable was not common during that part of the trip because of the larger group size and pre-assigned duties could be closely followed. Additionally since the journey was just starting military standards were more closely followed.

My personal experience as a long-term member of both the regular military and the reserves showed me some interesting differences in how discipline and military standards are applied. In a reserve force discipline and standards are less stringently adhered to whenever possible. Reserves are only part of the time military so they also bring their civilian side with them. The further from headquarters or high command the force is, and the longer the force is in the field the looser the standards and discipline become. (Bear in mind that the traditional customs and discipline are primarily designed to get the group to act as a team to get the mission completed.) The focus becomes more on getting the job done. (Whenever a reserve force is being visited or inspected by the regular forces the old standard is that the first order of business is to make sure everyone looks and acts military).

Any group that spends time together requires less discipline and standards for guidance the longer they are together. This becomes particularly true as the job becomes more difficult to perform. A good leader who has a well trained team will let that team do its job, only stepping in when necessary for guidance or correction.

One of the big problems Clark had that first winter at Camp Dubois was converting the Corps of Discovery from a group of individual civilians to a military team. He spent the winter training the men to become the kind of unit he needed to have. He knew he had a reserve force, at least in spirit, because a quarter of the men were civilian recruits. Lewis and Clark had a well trained team when they departed Fort Mandan. They had been trained well by Clark which was followed by a year of seasoning experience, with correction applied when necessary. But this force was a reserve force in its thinking and acting.

When the Corps left Mandan the next spring things started changing. They were a smaller group by about one third so pre-assigned duties were not so closely followed. Fewer people to do all the things that had to be done meant whoever was available did what needed to be done. This became very evident during the Expedition's return trip in 1806. By this time military protocol was relaxed since the Corps was split into smaller groups such as scouting parties, special work details, hunters, etc. One of the main cooks spent nearly the entire return trip as a hunter so had little time for cooking. As a result of these changes the Corps had a tendency to combine messes. They probably would form a single mess of whoever was in camp when several groups were gone.

The best methods of cooking when a party was on the move had been well developed and it was simple enough that any of the men could do the job. Add to that the fact that these men were well accustomed to frontier life so they knew how to cook well enough to at least feed themselves. By the time the Corps left Fort Clatsop there probably was little if any real need for a specifically designated cook for each of three separate messes. When it was time to eat one of the sergeants would task one, two or three people (depending upon the size of the party) to prepare the meal. This may well have been a task that fell to invalids who couldn't do other tasks, such as hunting.

Our testing has shown that one 5 gallon kettle can be used to cook enough to feed thirty men sufficient quantities of bread, red meat, fish, and birds and do so well within the allotted time. Combining messes allowed whoever was cooking to specialize thus making the chore easier. One person that knew how to prepare the complete meal could easily supervise two others that knew little about cooking without sacrificing time or results.

If one fire is all that is needed it makes no sense to build more. Why would the Corps go to the added work of cutting firewood, getting tripods, etc for three or four fires? We found from or testing that two cooks could easily feed the entire Corps in the allotted time if they combined messes and used a single fire.

During Lewis' trip from Travelers Rest to the Great Falls he had one of the cooks with him. That cook probably cooked for the entire party of ten. Clark had the other two cooks in his party of 23, but because hunters were out and others that were gone from time to time they probably had a single mess with two cooks working together. This concept continued when Ordway split from Clark at the Three Forks taking both cooks in his group. From the Three Forks downriver until the Expedition was reunited below the mouth of the Yellowstone some combination of York and Charbonneau probably cooked for Clark's party.

Once the Expedition was back into a single party the goal was to get downriver as quickly as possible so things got pretty loose with the attitude being to do only what needed to be done and as quickly as possible. Combined messes would have been the quickest and easiest way to feed the men.

The nature of the mission—they were exploring a region rather than fighting a battle—and the length of time they were in the field—two years rather than a few months—drew the men into a team that allowed the Captains to relax military protocol. We know the Corps varied from military standards so when combining messes made sense we are convinced the Expedition did so.

Chapter 20
Completing the Picture

We discovered through examining the journals that the Corps of Discovery actually had two different types of camps, a short camp of a few hours and a longer camp of several days or more. The Expedition's travel pattern seemed to be that they would travel hard for a number of days then stop for a time to rest, repair equipment, etc. These types of camps are in addition to the winters spent at Fort Mandan and Fort Clatsop, which we will consider separately.

Short Camps

Cooking in a short camp depended upon speed and simplicity. Each mess would build a small fire and use willows cut nearby to make a tripod to hold a brass kettle hanging over the fire. When the grease or lard in the kettle was hot supper was cooked in the oil. Supper would be strips of dried or freshly shot meat; fry bread made from flour or cornmeal; and maybe freshly caught fish or birds such as goose, ducks or quail. To top of the meal if in season maybe freshly picked berries that were eaten raw.

While the food was cooking in the oil if there was extra meat to be dried horizontal strips of willow would be fastened to the legs of the tripod and strips of meat would be hung over them to dry. The heat of the fire would help the heat from the sun to dry the meat faster. Drying meat meant drying the blood out of it. This basic preservation method enabled the meat to be kept for four to five days instead of two or three.

After the food was cooked the kettle of grease was set off the fire and left to cool and harden. Building the fire, heating the grease, cooking and eating the food took about an hour and a half. When the meal was over the boats were repacked and the Corps of Discovery proceeded on until dark when they stopped for the night to sleep.

The short camp established for cooking and eating supper would be minimally equipped. Only things necessary to prepare and eat the meal would be unpacked. No tents or other shelters were set up. If protection from the sun or wind was required camp would be located so trees or hills gave the necessary shelter.

Camp for the night was simply securing the boats at the riverbank then rolling out blankets and going to sleep for the night. The Captains' tent, or later their leather lodge, was normally erected and if the weather was bad some other shelters may be erected for the enlisted men.. There was no cooking in this camp that night or the next morning before the few personal items were reloaded and the boats pushed off, continuing upriver.

The journals frequently mention eating breakfast before departure. However this meal was seldom cooked. It was usually leftovers from supper the night before. If the Expedition was delayed in their departure and breakfast was cooked, it was much the same meal as had been cooked for supper the evening before and cooked in similar style.

Long Camps

When the Corps of Discovery stopped to rest for a few days or for longer periods, the cook fires took on a different look; the look of greater permanency, depending on the planned length of stay. The camp at the Lower Portage or the long camp in Idaho where the Expedition waited for the snow to melt in the Rockies probably looked quite long-term. When these camps were established shelters were erected and usually arranged according to army regulations (campfire patterns found during archaeological excavations at the Lower Portage Camp verify this). These camps would look much like what a person would imagine a camp to be; tents for the men's living quarters along with designated areas for cooking, various work details (animal butchering, hide preparation, baggage drying, boat repair, etc), latrines and guard posts.

In these camps time was not the primary factor influencing cooking technique so we would expect to see roasting on a spit, possibly some dutch oven cooking-stews, tarts and dumplings are all mentioned. Some other foods would be cooked that gave variation to the same limited foods eaten while the Expedition was traveling. Probably more fresh fish was eaten since there was time for some of the men to go fishing. At least some of these fish would have been pan fried in grease. Clark mentioned tarts several times in his notes, so this dessert is now a possibility. Vegetables such as squash were cooked, probably boiled in water. Rations of dried peas and beans brought from St. Louis could well have been prepared in these long camps. For example Lewis recorded the dinner for July 4, 1805 at the Upper Portage Camp as including bacon and beans.

To make this additional cooking possible a more permanent and versatile means of supporting the food over the cook fire would be needed. We are convinced that the cooks would not exclusively use willow tripods; wooden uprights and iron spits were erected over a larger pit so several things could be cooking at the same time.

On one end of the oval-shaped fire would be a roast on a spit; on the other end would be a tripod with a large kettle of beans slowly simmering away. The middle could be used for such things as frying fish by removing the lid from a dutch oven and using the flat-bottomed oven as a frying pan.

In these longer camps with more extensive cooking being done long handled utensils would have been necessary. Long handled forks to secure a roast on a spit and turn it for even cooking and for taking the cooked squash out of the hot water; a long handled spatula to turn the fish being fried' long handled spoons to stir the various beans, peas and stews that would be cooked.

While the Expedition was west of the continental divide, in present day Idaho, Washington and Oregon, they ate camas and cous roots extensively. The most common method for preparing these roots was to add water and make them into a mush. To add flavor grease was sometimes added or small pieces of meat. The meat was cut into small pieces then cooked over the fire, like making stew. When done it was added to the mush to greatly increase the flavor of the otherwise tasteless roots.

In the longer camps the messes may very well have been combined into a single cooking area. Since these cooks were not cooks by choice, but by order of the Captains, their skills in preparing a multi-course meal were not necessarily the greatest. With a single large cook fire each cook could oversee a certain area. By combining their efforts a more varied meal could be prepared.

Winter Camps

Cooking during the three winter camps would vary depending on the weather. A standard for the Army of that time would be to cook outside in a designated cooking area (which would take on an appearance much like that described above under long camps). However during foul weather; rain, snow, cold, etc., cooking was done in the men's quarters utilizing the fireplace. This was probably the same way the Corps of Discovery did theirs. At Camp Dubois they undoubtedly cooked in both areas since weather conditions were not as severe as during the winters at Mandan or Clatsop. The huts that served as quarters for the men were quite small so if weather permitted it may have been preferable to cook outside where there would be more room.

The fireplaces in the huts were small with no room to put a tripod over the fire to suspend a kettle from. The normal way to suspend a kettle would have been from a fireplace crane attached to one of the sidewalls of the fireplace. However there is no evidence this was used since they probably did not have enough iron with them to build these cranes or the right materials to fasten them into the rockwork of the chimney. They could have used an iron or wooden spit embedded in the rockwork of the chimney that was removable. Cooking kettles could then be suspended from the iron spit with an "S" hook Another possibility for cooking would have been a kettle on the hearth in front of the fire that simmered along for several hours as the soup or stew inside it cooked. Meat may well have been cut in small pieces then each man cooked their own by skewering a piece on a long handled fork and holding it to the fire much like today's camper may cook hot dogs. We are not told enough information to make a positive statement. We do know that the fireplaces were too small to put a tripod inside the fireplace to suspend the cooking kettle from.

Winter camps would have been an ideal place and time for using cast iron dutch ovens. Their versatility would have been welcome since breads and dumplings could have been included in the men's diet. Cast iron dissipates heat better than copper so a pot of beans or stew would cook more evenly with less risk of burning. Cooking at Mandan and Clatsop was probably done in the men's quarters most of the time to give protection from the weather. The winter at Mandan saw long periods of record cold while all but six days at Clatsop it rained.

Hunters and Small Groups

Another set of camps were those of hunters and other small groups such as the small party that went with Lewis on his overland trip from the mouth of the Marias to the Great Falls of the Missouri. When hunters went out for the day they carried with them their day's ration of food that had been cooked the night before, so they ate the same as if they were with the main party bringing the boats up the river. If they planned on being out overnight they may take an extra ration of food with them.

On several occasions hunters were unable to return to camp at night but had not planned on being away. The journal keepers give us the sense that the hunters would simply improvise by killing an animal and cooking the meat for their supper and breakfast. They carried no particular cooking equipment so they would probably cut the meat into small pieces and roast it over a fire like we would roast hot dogs. If their game was a small animal such as a rabbit or grouse they may improvise a spit with green tree branches and roast it. These camps are the type that most commonly come to mind when we think of mountain men or other adventurers.

Although Lewis makes no mention of taking any cooking equipment when he went on his trek to the upper Marias (Camp Disappointment) in July of 1806 his journal references frying a trout and eating the last of the Expedition's supply of cous roots. He said he made a mush and added a little bit of bear oil. From these hints we realize he must have had at least a small kettle to mix the dried root with water in to make the mush. That same kettle could have been used to fry the fish in a make-do situation.

When small parties were traveling by horseback or canoe they likely carried some cooking equipment such as a small brass kettle. However if they were on foot it becomes less likely, although they would have their military experience to draw upon in selecting what they carried. We do have some reports in the journals of small parties carrying kettles with them. The primary determiner was how long they were to be gone from the main group. If they were just out hunting for the day they took food with them that was already cooked. If they planned to be gone for a while they took some cooking equipment.

The men of the Corps of Discovery were very familiar with these individual or small group situations (refer to the earlier discussion in chapter 5 about hunting pioneers). Their cooking methods were the basic survivalist type practiced uncountable times prior to the journey.

Typical cooking fire set up for a camp when the Corps stopped for a few days. When they only stopped long enough to eat supper the main difference would be less equipment unloaded and they normally would not have a tent erected.

Chapter 21
In a Nutshell

At the start of this book the question was asked, "What did the Expedition eat; how were the foods prepared; and what equipment did the cooks use to prepare it."

We have traced through the records including the journals written during the journey along with other primary documents prepared just before and just after the Expedition to find very little was recorded about preparation methods or equipment. The few times cooking methods or equipment was referred to was when they were unusual or in combination with situations that were worthy of recording, such as a new way of food preparation or a new food such as camas or cous roots was discussed. Most of the equipment references are deduced from the food they mention eating; Lewis comments that they caught a trout and fried it in a little bear oil. We were able to develop a reasonably complete list of foods eaten and where they came from, but all else was quite sketchy.

To fill in the gaps in the documented record created by journal keepers who didn't record the common activities or those that they deemed unimportant we turned to secondary documents and field testing. As a result we were able to make logical conclusions and develop several theories which gave us a pretty good picture of what these people ate, where they got the food, who cooked, how they cooked, what equipment was used and what a person who "walked into the Corps camp" would likely expect to find. Along the way we were able to dispel several commonly heard myths that cleared some of the confusion surrounding the Expeditions food and cooking habits.

We floundered for direction a good bit until we realized that the keys to the Expedition's cooking and eating habits were the element of time and simplicity and the fact that food did not play the same role in their lives as it does ours today. They viewed it as basic for survival and were very unconcerned about variety and the myriad of attributes we give food today. They merely wanted something to put in their bellies to fill them up so they could continue their work at hand.

I can't overemphasize the value of our field testing to fill the gaps in the recorded data. These tests can be favorably compared to laboratory experiments since situations were duplicated and the variables were controlled to find the outcomes as each single variable was allowed to vary.

As we worked through testing various possibilities we eventually came upon the methods that showed themselves to be the most probable. To determine "most probable" we had to consider the mindset of the Corps-military and civilian; background, knowledge and abilities; availability of required materials needed; ease of performing tasks; frequency of task done during the journey; time required to complete and time available; number of people involved-are you feeding one or twenty.

Our job was made more difficult, and probably much more controversial, because there could be several ways of doing the related tasks or methods for cooking. Similarly, there could be several different kinds of equipment used in the preparation. We had to test and try to determine what were the most likely methods and equipment the Corps used. That is what we have presented in this book.

During the seasons that plants were edible or in locations of abundant game animals people who live a subsistence lifestyle eat plentifully. At other times they vary from adequate to barely surviving. Death from starvation is not uncommon. For this lifestyle great amounts of time are spent in the daily quest for food. We see all of these facets of life in the journals of the Corps of Discovery's trek.

As the exploring party worked its way upriver from St. Louis to Fort Mandan the season was right to harvest substantial quantities of plants. Therefore their diet was rich and varied. As fall turned to winter then spring they had a much greater dependence on meat from game animals they may discover.

Winter in the Rocky Mountains and a totally different environment to the west reduced the Corps to a basic survival lifestyle. Their diet consisted in a large part of foods they were unaccustomed to and their bodies reacted accordingly. Only the very best hunters were able to take what red meat was available. The Corps survived only because of the planning, preparation and training that preceded the journey.

One of the more intriguing elements of the planning process was for a military unit to "extend the hand of peace" to a traditional enemy—Indians. (While this may be a routine role for the military today, it was something new for the American army of 1800) Jefferson was obviously looking at the prospects of establishing trade by telling the Corps of Discovery to trade (for food) and establish peace throughout the region they would travel. But with the Expedition's food source depending on their treating the natives more as equals than as enemies they made large strides in staving off starvation during much of the journey. Several tribes made substantial food gifts to the Corps.

The Expedition left many volumes of records that help us understand their work. We complain about the absence of detail in some aspects of what they wrote or what may have been left out. However the largest hole in documenting the Lewis and Clark Expedition remains the lack of information on how the members of the Corps of Discovery were selected. How the Captains were able to get people that could be formed into a cohesive team, could be trained as the journey required, and had the abilities needed to survive and complete the job at hand remains a mystery.

Scenes from a re-enactment of the 1805 portage around the Great Falls of the Missouri. It was a steep and difficult climb out of the river bottom onto the fairly level prairie; Clark said they had to pull with all their strength just to get the canoes up to where they were then loaded with baggage and hauled 15 miles across the prairie. In 1806 the carts were pulled by horses.

Appendix I: Recipes in the Journals

Boudin Blanc

On May 9, 1805 Lewis describes how boudin blanc, or white pudding is made: From the buffalo I killed we saved the necessary materials for making what our right-hand cook, Charbonneau, calls the boudin blanc. About 6 feet of the lower extremity of the large gut of the buffalo is the first morsel that the cook makes love to, this he holds fast at one end with the right hand while with the forefinger and thumb of the left he gently compresses it and discharges what he says is not good to eat, but of which in the equal we get a moderate portion. The muscle lying underneath the shoulder blade next to the back and fillets are next sought. These are needed up very fine with a good portion of kidney suet. To this composition is then added a just proportion of pepper and salt and a small quantity of flour. Thus far advanced our skillful operator seizes his receptacle, which has never once touched the water-for that would entirely destroy the regular order of the whole procedure. You will not forget that the side you now see is that covered with a good coat of fat provided the animal be in good order. The operator seizes the receptacle I say and tying it fast at one end turns it inwards and begins now with repeated evolutions of the hand and arm and a brisk motion of the finger and thumb to put in what he says is "bon pour manger". Thus by stuffing and compressing he soon distends the receptacle to the utmost limits of its power of expansion, and in the course of the operation its longitudinal progress it drives from the other end of the receptacle a much larger portion of discharge than was previously discharged by the finger and thumb of the left hand in a former part of the operation. Thus when the sides of the receptacle are skillfully exchanged the outer for the inner and all is completely filled with something good to eat, it is tied at the other end, but not any cut off for that would make the pattern too scant. It is then baptized in the Missouri with two dips and a flirt and bobbed into a kettle from whence after it be well boiled, it is taken and fried with bears oil until it becomes brown...

Nez Perce Method for Cooking Grizzly Bear

After establishing their camp near the Nez Perce in June of 1806 to wait for the snow in the Rockies to melt the Expedition's hunters were successful in killing two grizzly bears. They shared their good fortunes with about 15 Nez Perce who were close by. To the Indians this was somewhat of a treat since they did not get bear meat very often. Captain Lewis described how the Indians cooked the meat.

"(They) prepared a brisk fire of dry wood on which they threw a parsel of smooth stones from the river. When the fire had burned down and heated the stones, they placed them level and laid on a parsel of pine boughs. On these they laid the flesh of the bear in flitches (layers) placing boughs between each course of meat and then covering it thickly with pine boughs. After this they poured on a small quantity of water and covered the whole over with earth to a depth of four inches. In this situation they suffered it to remain about three hours when they took it out."

"I tasted this meat and found it much more tender than that which we had roasted or boiled, but the strong flavor of pine destroyed it for my palate."

Quawmash Roots

On June 11, 1806 Lewis describes how the Nez Perce cook Quawmash roots. We know this plant to be camas roots, which was commonly used by most of the Indians in the northwest region.

...when they have collected a considerable quantity of these roots or 20 to 30 bushels which they readily do by means of a stick sharpened on one end, they dig away the surface of the earth forming a circular concavity of 2 ½ feet in the center and 10 feet in diameter. They next collect a parsel of split dry wood with which they cover this basin in the ground perhaps a foot thick, they next collect a large parsel of stones of about 4 or 6 lbs weight which are placed on the dry wood; fire is then set to the wood which burning heats the stones; when the fire has subsided and the stones are sufficiently heated which are nearly red heat, they are adjusted in such manner in the whole as to form as level a surface as possible, a small quantity of earth is sprinkled over the stones and a layer of grass about an inch thick is put over the stones; the roots, which have been previously divested of the black or outer coat and radicles which rub off easily with the fingers, are now laid on in a conical pile, are then covered with a layer of grass about 2 or 3 inches thick; water is now thrown on the summit of the pile and passes through the roots and to the hot stones at the bottom; some water is also poured around the edges of the hole and also finds its way to the hot stones; as soon as they discover from the quantity of steam which issues that the water has found its way generally to the hot stones, they cover the roots and grass over with earth to the depth of four inches and then build a fire of dry wood all over the conical mound which they continue to renew through the course of the night for ten or 12 hours after which it is suffered to cool two or three hours when the earth and grass are removed and the roots thus sweated and cooked with steam are taken out, and most commonly exposed to the sun on scaffolds until they become dry, when they are black and of a sweet agreeable flavor. These roots are fit for use when first taken from the pit, are soft of a sweetish taste and much the consistency of a roasted onion; but if they are suffered to remain in bulk 24 hours after being cooked they spoil. If the design is to make bread or cakes of these roots they undergo a second process of baking being previously pounded after the first baking between two stones until they are reduced to the consistency of dough and then rolled in grass in cakes of eight or ten lbs are returned to the sweat intermixed with fresh roots in order that the steam may get freely to these loaves of bread. When taken out the second time the women make up this dough into cakes of various shapes and sizes usually from ½ to ¾ of an inch thick and expose it on sticks to dry in the sun, or place it over the smoke of their fires. The bread this prepared if kept free from moisture will keep sound for a great length of time. This bread or the dryed roots are frequently eaten alone by the natives without further preparation, and when they have them in abundance they form an ingredient in almost every dish they prepare. This root is palateable but disagrees with me in every shape I have ever used it.

Appendix II: From the Corps Orderly Book

Detachment Order April 1, 1804

    The Commanding Officers did yesterday proceed to take the necessary enlistments and select the Detachment destined for the Expedition through the interior of the Continent of North America; and have accordingly selected the persons herein after mentioned, as those which are to constitute their Permanent Detachment. *[names listed]*

    The Commanding Officers do also retain in their service until further orders the following persons: Richard Warvington, Robert Frasure, John Robertson and John Boyley who whilst they remain with the Detachment shall be incorporated with the Second and Third Squads of the same and are to be treated in all respects as those men who form the Permanent Detachment except with regard to an advance of pay and the distribution of arms and accoutrements intended for the Expedition.

    The following persons: Charles Floyd, John Ordway and Nathaniel Pryor are this day appointed Sergeants with equal power unless when otherwise specially ordered. The authority, pay and emoluments attached to the said rank of Sergeants in the Military Service of the United States and to hold the said appointments and be respected accordingly during their good behavior or the will and pleasure of the said Commanding Officers.

    To insure order among the party, as well as to promote a regular police in camp, the Commanding Officers, have thought to divide the detachment into three squads and to place a Sergeant in Command of each, who are held immediately responsible to the Commanding Officers for the regular and orderly deportment of the individuals composing their respective squads.

    The following individuals after being duly balloted for, have fallen in the several squads as hereinafter stated and are accordingly placed under the direction of the Sergeants whose names proceeds those of his squad. *[names listed]*

    The camp kettles and other public utensils for cooking shall be produced this evening after the parade is dismissed and an equal division shall take place of the same among the non commissioned officers commanding the squads. Those non commissioned officers shall make an equal division of the proportion of those utensils between their own messes of their respective squads. Each squad shall be divided into two messes at the head of one mess which the commanding sergeant shall preside. The sergeants messes will consist of four privates only to be admitted under his discretion, the balance of each squad shall form the second mess of each squad.

    During the indisposition of Sergeant Pryor, George Shannon is appointed to discharge the said Pryor's duty in his squad.

    The party for the convenience of being more immediately under the eye of the several sergeants having charge of them will make the necessary exchange of their bunks and rooms for that purpose as shall be verbally directed by us.

    Until otherwise directed Sergeant John Ordway will continue to keep the roster and detail the men of the detachment for the several duties which it may be necessary they should perform, and also to transcribe in a book furnished him for that purpose those or such other orders as the Commanding Officers shall think proper to publish from time to time for the government of the party.

                                                  Signed
                                                Meriwether Lewis
                                                Wm. Clark

Detachment Order May 26, 1804

The Commanding Officers direct that the three squads under the command of Sergts Floyd, Ordway and Pryor heretofore forming two messes each, shall until further orders constitute three messes only, the same being altered and organized as follows:
[members of each mess named]
The Commanding Officers further direct that the remainder of the detachment shall form two messes and that the same be constituted as follows:
[engages named to one mess;
soldiers to return named to second mess]

This order continued by assigning men to boats and giving sergeants and their squads their daily duties. What follows is the part of the duties that apply to the daily ration of food:

...provision for one day will be issued to the party on each evening after we have encamped. The same will be cooked on that evening by the several messes and a proportion of it reserved for the next day as no cooking will be allowed in the day while on the march.

Sergt John Ordway will continue to issue the provisions and make the details for guard and other duty. The day after tomorrow lyed corn and grease will be issued to the party, the next day pork and flour, and the day following Indian meal and pork, and in conformity to that routine provisions will continue to be issued to the party until further orders. Should any of the messes prefer Indian meal to flour they may receive it accordingly. No pork is to be issued when we have fresh meat on hand.

Meriwether Lewis Capt
Wm Clark Capt

Detachment Order July 8, 1804

In order to insure a prudent and regular use of all provisions issued to the crew of the bateaux in future as also to provide for the equal distribution of the same among the individuals of the several messes, the Commanding Officers do appoint the following persons to receive, cook and take charge of provisions which may from time to time be issued to their respective messes; John Thompson to Sergt Floyd's mess, William Warner to Sergt Ordway's mess, and John Collins to Sergt Pryor's mess. These Superintendents of Provisions are held immediately responsible to the Commanding Officers for a judicious consumption of the provisions which they receive. They are to cook the same for their several messes in due time and in such manner as is most wholesome and best calculated to afford the greatest proportion of nutriment. In their mode of cooking they are to exercise their own judgment. They shall also point out what part and what proportion of the mess provisions are to be consumed at each stated meal, morning, noon and night. Nor is any man at any time to take or consume any part of the mess provisions without the privity, knowledge and consent of the Superintendent. The Superintendent is also held responsible for all the cooking utensils of his mess.

In consideration of the duties imposed by this order on Thompson, Warner and Collins, they will in future be exempt from guard duty though they will still be held on the roster for that duty and their regular tour shall be performed by someone else of their respective messes. They are exempted also from pitching tents of the mess, collecting firewood, and fork poles etc for cooking and drying such fresh meat as may be furnished them. Those duties are to be also performed by the other members of the mess.

                                                   M Lewis
                                                   Wm Clark

[Pvt Peter Weiser was named to replace Pvt John Thompson as Superintendent of Provisions for Sergt Floyd's mess on August 12, 1804. This was a verbal command with no reason why given.]

Detachment Order August 28, 1804

The Commanding Officers direct that the two messes who form the crews of the pirogues shall select each one man from their mess for the purpose of cooking and that these cooks as well as those previously appointed to the messes of the barge crew shall in future be exempted from mounting guard or any detail for that duty. They are therefore no longer to be held on the roster.

> M. Lewis Capt.
> 1st U'S. Regt. Infty.
> Wm. Clark Cpt. &

[The two crews of the pirogues were the engages and the soldiers who would be returning to St. Louis. The barge crew was the permanent party that would go to the Pacific.]

Appendix III: Foods Eaten by Corps of Discovery

Collect	Acorn (white oak)	
Collect	Lotus Lily	*Nelumbo lutea*
Collect	Antelope	
Collect	Apples	
Collect	Bear	
Collect	Beaver	
Collect	Bighorn Sheep	
Collect	Black berry	*Rubus ursinus*
Collect	Blue currant (black, golden)	*Ribes americanum, Ribes aureum*
Collect	Brant	
Collect	Buffalo	
Collect	Buffalo berry	*Shepherdia argentea*
Collect	Burrowing Squirrel	
Collect	Cherries	
Collect	Chokecherries	*Prunus virginiana*
Collect	Coyote	
Collect	Deer	
Collect	Duck	
Collect	Elk	
Collect	Fish (fresh)	
Collect	Geese	
Collect	Gooseberry	*Ribes divaricatum*
Collect	Grapes (Raccoon) Shiny Oregon Grape Dull Oregon Grape	*Ampelopsis cordata Berberis aquifolium Berberis nervosa*
Collect	Greens (water cress)	*Rorippa nasturtium-aquaticum*
Collect	Grouse	
Collect	Hazel nuts	*Corylus americana*
Collect	High Bush American Cranberry	*Viburnum trilobum*
Collect	Onions	
Collect	Otter	
Collect	Pawpaw	*Asimina triloba*
Collect	Pelican	
Collect	Pheasant (Grouse)	
Collect	Pigeons	
Collect	Plums (Ground)	*Astragalus crassicarpus*
Collect	Prairie Dog	
Collect	Quail	

Collect	Rabbits	
Collect	Raccoon	
Collect	Raspberries (Salmonberry)	*Rubus spectabilis*
Collect	Hawthorn (Black)	*Crataegus douglasii*
Collect	Service berries	*Amelanchier alnifolia*
Collect	Soft shell turtle	
Collect	Squirrel	
Collect	Strawberries	*Fragaria glauca*
Collect	Turkey	
Collect	Wild onions (Geyer's onion)	*Allium geyeri*
Others	Chicken	
Others	Chocolate	
Others	Eggs	
Others	Milk	
Sacajawea	Breadroot (white apple, Indian turnip)	*Psoralea esculenta*
Sacajawea	Indian Potatoes (Western Spring Beauty)	*Claytonia lanceolata*
Sacajawea	Yampah (Indian carrot) or (Fennel)	*Perideridia gairdneri*
Taken	Bacon	
Taken	Beans	
Taken	Beef	
Taken	Bisquits	
Taken	Bread	
Taken	Butter	
Taken	Cheese	
Taken	Coffee	
Taken	Corn (meal, parched)	
Taken	Flour	
Taken	Honey	
Taken	Lard	
Taken	Peas	
Taken	Pork	
Taken	Portable soup	
Taken	Potatoes	
Taken	Salt	
Taken	Sugar	
Taken	Turnips	
Taken	Vinegar	
Trade	Anchovies	
Trade	Bitterroot	*Lewisia rediviva*
Trade	Camas root (quawmash)	*Camassia quamash*
Trade	Cattail	*Typha latifolia*
Trade	Cous root	*Lomatium cous*
Trade	Dog	
Trade	Fish (dried)	

Trade	Giant Horsetail Rush	*Equisetum talmateia*
Trade	Horse	
Trade	Licorice	*Glycyrrhiza lepidota*
Trade	Lily, mariposa	*Calochortus elegans*
Trade	Lily, white water	*Nymphaea sp*
Trade	Sunflower	*Helianthus annuus*
Trade	Squash/vegetables	
Trade	Thistle (edible))	*Cirsium edule*
Trade	Tobacco Root	*Valeriana edulis*
Trade	Wapato (Arrowhead)	*Sagittarria latifolia*
Trade	Watermelon	
Trade	Western bracken fern	*Pteridium aquilinum*
Trade	Whale blubber	
Trade	Wild hyacinth	*Triteleia grandiflora*

Notes:

Sources of Foods:
Taken = Initial food supply taken when departing Camp Dubois
Collect = Collected by the Corps through their living off the land
Trade = Obtained by trading with Indians along the route
Sacajawea = Collected by Sacajawea along the route
Others = Obtained from traders coming upriver or at settlements on the lower Missouri

Appendix IV: How it was Cooked

Red meat, birds	-boiled most of the time either in oil or water -sometimes roasted -rabbit, squirrel, prairie dog, raccoon where probably roasted
Fish	-boiled in oil or water -mention made of pan frying and roasting
Flour, cornmeal	-primarily made into bread and boiled in oil or fried -cornmeal also made into mush with a little grease or meat for flavoring -flour also used for dumplings, tarts, cobblers
Fruits	-eaten raw -possibly some used in tarts, cobblers
Vegetables (turnip, bread root, Yampah, squash)	-boiled in water
Whale blubber	-rendered for cooking oil -boiled in water and eaten as meat
Eggs	-probably hard boiled
Portable soup	-mixed with water and heated -possibly added to stew or soup at Fort Mandan
Roots (camas, cous, cattail, rush, licorice, lily, sunflower, thistle, tobacco root, wapato, fern, hyacinth, bitterroot)	-mostly purchased from Indians so the were already cooked/dried (see appendix I, quawmash roots). To eat they were mixed with water to make mush -they could be eaten without further preparation -sometimes ground into flour and used to make bread

Appendix V: Pocket Soup, Portable Soup, Beefe Glue

Several cookbooks have talked about making portable soup. They all produce a soup that turns out to be quite tasty. However, the portable soup that the Corps of Discovery took with them was reported by the men as being horrible; the last thing to be eaten before starvation. Consequently we believe that these cookbooks are not replicating what was taken on the Expedition. We have included this to give the reader an idea of what portable soup really was. The first paragraph is the best description of what the Corps of Discovery had with them.

The following discussion of portable soup was taken from Bob's Black Powder Notebook, Bob Spencer web page
http://members.aye.net/~bspen/trailfoods.html

In the literature of colonial times, mention is made of "pocket soup", "portable soup", "powdered beefe" and "glue broth". This was their equivalent of bullion, and they used it in about the same way. Basically, it is a simple broth made by boiling bones and meat trimmings for several hours until it concentrates into a syrupy consistency, then pouring it in a shallow layer and allowing it to dry till hard (by leaving it in the sun for several days). This was reconstituted by dissolving it in hot water, and served as a sort of instant stock.

Here's my adaptation of the old recipes. This presumes a freshly killed, butchered and de-boned deer.

Break all the long bones to allow access to the marrow, add most of the spine and neck bones, whatever tendons or ligaments are available and all the joints on both front and back legs. Put these into a large stock pot, cover them with water and bring it to a boil. After boiling covered for 8 hours, remove all the meat and bones and let the liquid cool overnight in the refrigerator. In the morning there will a layer of hard white fat and floating solids on top, which you should scoop off completely. Continued to boil, now uncovered, and in about 6 more hours the broth will be very thick, like warm molasses. Be careful not to let it burn. Pour this into a shallow bread pan, and it will make a layer about 3/8 inch thick. When it cools, it will be the consistency of hard jello, very rubbery. Remove it from the pan, cut it into rectangles, and place it on waxed paper then dry it in the dehydrator for one day, and it will turn hard. It tastes like a very flavorful meat broth. An old recipe calls for salt, pepper, mace, cloves and brandy to be added during the boiling, which sounds much more highly seasoned than mine. I add only 1 teaspoon of salt.

It is necessary to have an adequate amount of cartilage and connective tissue to begin with, because the "soup" won't ever get hard without it, that's where the gelatin comes from. If you include all the joints and the spine, it should be alright, but you can add a couple of pounds of pig's feet to make sure. I've never had to do that.

One of the ways the long hunters used this food was to reconstitute it by boiling, then adding a few spoonfuls of parched corn and cooking until this thickened. The result is a sort of meat flavored porridge, very tasty and satisfying.

A square of pocket soup about 2 X 3 inches dissolved in 1 1/2 cup boiling water makes a nice stock. I usually add 1/4 cup toasted yellow stone ground corn meal to the broth and simmer it for about 10 minutes until it thickens nicely. Salt and pepper seasoning make a very tasty meat-flavored "hasty pudding". This is easy to do in camp, and carrying enough cornmeal and pocket soup squares for several days' meals is very little trouble, both being so light and compact. Jerky cut into small pieces and added to the mix makes an even more sustaining meal, and adds little to the weight of my pack.

Chapter End Notes

Chapter 2: To Camp Dubois
1. DeVoto, Bernard, editor, The Journals of Lewis and Clark (Boston, Houghton Mifflin Company, 1953) page 484.
2. Moore, Robert J. Jr. and Haynes, Michael, Lewis and Clark; Tailor Made, Trail Worn (Helena, Far Country Press, 2003) page 38.
3. Moore, page 38.
4. DeVoto, page 484.
5. Jackson, Donald, editor, Letters of the Lewis and Clark Expedition (Urbana, University of Illinois Press, 1962) page 76.
6. Ibid page 117.
7. Moore, page 38.
8. Moore, page 38.
9. Moore, page 37.
10. Moore, page 37.
11. Moore, page 37.
12. Elting, John, Col, US Army, ret. American Army Life (New York, Charles Scribner & Sons, 1982) page 59.
13. Jackson, page 76.

Chapter 3: Leaving St. Charles...up the Missouri
1. Elting, page 77.
2. Beare, Patricia and Myers, Judith. Adult Health Nursing, (the C.V. Mosley Co, 1990) page 1792.
3. Ambrose, Stephen, Undaunted Courage (New York, Simon & Schuster, 1995) page 148.

Chapter 4: Entering Indian Territory
1. Moulton, Gary E. editor, The Journals of the Lewis and Clark Expedition, vol 2 (Lincoln, University of Nebraska Press, 1983) page 496.
2. Moulton, vol 3, page 12.
3. Moulton, vol 3, page 854.
4. Ronda, James, Lewis and Clark Among the Indians (Lincoln, University of Nebraska Press, 1984) page 28.
5. Moulton, vol 9, page 69.
6. Moulton, vol 3, page 148.
7. Moulton, vol 3, page 160.
8. Moulton, vol 3, page 163.

Chapter 5: Winter at Mandan
1. Moulton, vol 4, page 489.
2. Moulton, vol 11, page 109.
3. Moulton, vol 9, page 117.

Chapter 6: Mandan to the Marias
1. Moulton, vol 3, page 327.
2. Moulton, vol 4, page 48.
3. Moulton, vol 4, page 100.
4. Moulton, vol 4, page 102.

5. Moulton, vol 4, page 120.
6. Moulton, vol 4, page 126.
7. Moulton, vol 4, page 141.
8. Moulton, vol 4, page 226.

Chapter 7: Marias River to Canoe Camp
1, Moulton, vol 4, page 274.
2. Moulton, vol 4, page 279.
3. Moulton, vol 4, page 362.
4. Moulton, vol 10, page 109.
5. Moulton, vol 4, page 379.

Chapter 8: Canoe Camp to Beaverhead Rock
1. Moulton, vol 5, page 18.
2. Moulton, vol 11, page 255.
3. Moulton, vol 11, page 263.

Chapter 9: Beaverhead Rock to the Nez Perces
1. Moulton, vol 11, page 265.
2. Moulton, vol 5, page 74.
3. Moulton, vol 5, page 74.
4. Moulton, vol 5, page 83.
5. Moulton, vol 5, page 88.
6. Moulton, vol 5, page 105.
7. Moulton, vol 5, page 144.
8. Moulton, vol 5, page 171.
9. Moulton, vol 5, page 177.
10. Moulton, vol 9, page 217.
11. Moulton, vol 9, page 223.
12. Moulton, vol 11, page 319.
13. Moulton, vol 11, page 319-320.
14. Moulton, vol 10, page 144-145.
15. Moulton, vol 9, page 227.
16. Moulton, vol 5, page 222-223.
17. Moulton, vol 5, page 226.

Chapter 10: Nez Perces to the Pacific Ocean
1. Moulton, vol 5, page 232.
2. Whitney and Rolfes. Understanding Nutrition, (Belmont, CA, Wadsworth/Thompson Learning, 2002) page 647
3. Website www.mayoclinic.com/health/food-poisoning/DS00981/DSECTION=symptoms
4. Moulton, vol 10, page 150.
5. Moulton, vol 10, page 152.
6. Moulton, vol 11, page 344.
7. Moulton, vol 5, page 288.
8. Moulton, vol 5, page 318.
9. Moulton, vol 5, page 341-342.
10. Moulton, vol 10, page 162.
11. Moulton, vol 5, page 363.

12. Moulton, vol 10, page 166.
13. Moulton, vol 6, page 33.
14. Moulton, vol 6, page 39-40.
15. Moulton, vol 6, page 42.
16. Moulton, vol 11, page 392.

Chapter 11: Winter on the Pacific Coast
1. Moulton, vol 11, page 171.
2. Moulton, vol 6, page 69.
3. Moulton, vol 11, page 398.
4. Moulton, vol 6, page 91.
5. Moulton, vol 6, page 93.
6. Moulton, vol 6, page 107.
7. Moulton, vol 6, page 118.
8. Moulton, vol 6, page 126.
9. Moulton, vol 6, page 135.
10. Website www.wedlinydomowe.com/smoking-meat.htm. Visit this site for more detailed discussion on this process.
11. Moulton, vol 6, page 138.
12. Moulton, vol 6, page 145.
13. Website www.arkson.com/health/vbone.htm
14. Moulton, vol 6, page 162.
15. Moulton, vol 6, page 166.
16. Moulton, vol 6, page 186.
17. Moulton, vol 6, page 200.
18. Moulton, vol 6, page 211.
19. Moulton, vol 6, page 215-216.
20. Moulton, vol 6, page 245.
21. Moulton, vol 6, page 193.
22. Moulton, vol 6, page 336.
23. Moulton, vol 6, page 378.
24. Moulton, vol 6, page 407.
25. Moulton, vol 9, page 278.

Chapter 12: Fort Clatsop to Travelers Rest
1. Moulton, vo 7, page 13.
2. Moulton, vol 7, page 34.
3. Moulton, vol 7, page 49-50.
4. Moulton, vol 7, page 88.
5. Moulton, vol 7, page 115.
6. Moulton, vol 7, page 142.
7. Moulton, vol 7, page 204.
8. Moulton, vol 7, page 227.
9. Moulton, vol 7, page 234.
10. Moulton, vol 7, page 265.
11. Moulton, vol 8, page 10.
12. Moulton, vol 8, page 23.
13. Moulton, vol 8, page 57.

Chapter 13: Travelers Rest and Onward
1. Moulton, vol 8, page 125.
2. Moulton, vol 8, page 133-135.
3. Moulton, vol 8, page 145.
4. Moulton, vol 8, page 146.
5. Moulton, vol 8, page 288.
6. Moulton, vol 8, page 366.

Chapter 14: Diet During the Journey
1. Kershaw, Linda. Edible and Medicinal Plants of the Rockies, (Edmonton, Alb, Canada, Lone Pine Publishing, 2000) page 105.

Chapter 15: Assembling the Data
1. Personal email from Gene Hickman 11/14/03 and 2/15/05. Also personal correspondence with Kenneth Wilk 11/17/03 and 1/26/04.
2. Moore, page 38.
3. Website www.arbuckles.com.
4. Ragsdale, John G. Dutch Ovens Chronicled; their use in the United States, (Fayetteville, University of Arkansas Press, 1991) page 11.
5. Website www.oldandsold.com.
6. Website http://oregontrail.blm.gov.
7. Website www.nwta.com Pvt Joseph Martin, Connecticut Militia 1776, memoirs.
8. Website www.nwta.com General Orders, Washington's Army 19 June 1778.
9. Website www.nwta.com .
10. Mulholland, James. A History of Metals in Colonial America (Fayetteville, University of Arkansas Press, 1981) page 116.
11. Mulholland, page 46.
12. Website www.oldandsold.com.
13. Moulton, vol 2 page 133.
14. Moulton, vol 2 page 175.

Chapter 16: Myths Exposed
1. Holden, Robert. Hunting Pioneers (Bowie, Heritage Books, 2000) page 2.
2. Martin, Paul and Szuter, Christine. "War Zones and Game Sinks in Lewis and Clark's West" Conservation Biology, February 1999, Volume 13, Number 1, pages 36-45. Paul Martin devised a basic "ration unit" that made it possible to total pounds of meat for various sized animals. He used Clark's August 22, 1805 journal entry that said it required one buffalo or four deer or two elk and two deer to feed the men for a day. Other animals were compared to the weights represented by these animals to create a formula of required meat. Ken Walcheck also used this concept in some of his writings on the subject.
3. Here is a sampling of such websites: www.dutchovenmagic.com, www.alpinetrail.com, www.dutchovencookware.com, http://home.earthlink.net/~rfrmac/Pages/dutch_info-2.html
4. Ragsdale, page 14.

5. Website www.osv.org. article written by Frank G. White, January 1980. This is the website of the Old Sturbridge Village.
6. Triber, Jayne E., A True Republican; the life of Paul Revere (Boston, University of Massachusetts Press, 1998) page 154.
7. Mulholland, page 31.
8. Mulholland, pages 55-58.
9. Leehey, Patrick M. Paul Revere-Artisan, Businessman and Patriot (Boston, Paul Revere Memorial Association, 1988) page 31.
10. Leehey, page 111.
11. Personal correspondence with Patrick Leehey 2/23/05 and 2/25/05.
12. Ernay, Renee Lynn. "The Revere Furnace 1787-1800" The Revere House Gazette, Issue No 26, Spring 1991 and Ernay, Renee Lynn. "The Revere Furnace 1787-1800" Unpublished Master's Thesis, University of Delaware, 1989.

Chapter 18: Weighing the Evidence
1. Bakeless, John. Lewis and Clark Partners in Discovery (New York, William Morrow & Co, 1947) page 36.
2. Macomb, Alexander. A Treatise on Martial Law and Courts Martial as Practices in the United States of America (Charleston, SC, US Military Philosophical Society, 1809) page 35.
3. Carstens, Kenneth C., editor. Calendar & Quartermaster Book of Ft. Jefferson, Kentucky 1780-1781 (Bowie, MD, Heritage Press, 2000) Part II, pages 27 & 49.
4. Jackson, vol 1, page 103.
5. Moore, page 25.
6. Personal interview with Colleen Sloan at Great Falls, Montana on January 31, 2004.

Sources

Ambrose, Stephen E, *Undaunted Courage*, New York, Simon and Schuster, 1996

Bakeless, John, *Lewis and Clark, Partners in Discovery*, New York, William Morrow & Company, 1947

Beare, Patricia and Myers, Judith, *Adult Health Nursing*, the C.V. Mosley Co, 1990

Burroughs, Raymond, "Game Trails of the Lewis and Clark Expedition," unpublished manuscript, Lewis and Clark Trail Heritage Foundation archives, Great Falls, MT, September 2, 1999

Carstens, Kenneth C., editor, *The Calendar and Quartermasters Books of General George Rogers Clark's Fort Jefferson, Kentucky 1780-1781*, Bowie, MD, Heritage Books, Inc., 2000

Cutright, Paul Russell, *Lewis and Clark Pioneering Naturalists*, Urbana, University of Illinois Press, 1969

DeVoto, Barnard, editor, *The Journals of Lewis and Clark*, Boston, Houghton Mifflin Company, 1953

Elting, John, Col, Us Army, ret. *American Army Life*, New York, Charles Scribner & Sons, 1982

Ernay, Renee Lynn, "The Revere Furnace 1787-1800," *The Revere House Gazette*, Issue No. 26, Spring 1991

Ernay, Renee Lynn, "The Revere Furnace 1787-1800" Master's Thesis, University of Delaware, 1989

Gibson, James B., *Imperial Russia in Frontier America*, New York, Oxford University Press, 1976

Hart, Jeff. *Montana Native Plants and Early People*, Helena, Montana Historical Society Press, 1992

Hickman, Gene Colonel, United States Air Force, retired, military historian. Personal correspondence November 14, 2003 and February 15, 2005.

Holden, Robert John, *The Hunting Pioneers*, Bowie, MD, Heritage Books, Inc., 2000

Jackson, Donald, editor. *Letters of the Lewis and Clark Expedition*, Urbania, University of Illinois Press, 1978.

Kershaw, Linda. *Edible and Medicinal Plants of the Rockies*, Edmonton, Alb, Canada, Lone Pine Publishing, 2000.

Leehey, Patrick M., *Paul Revere-Artisan, Businessman and Patriot*, Boston, Paul Revere Memorial Association, 1988

Leehey, Patrick M, Research Director at Paul Revere Memorial Association, Boston, Massachusetts. Personal correspondence February 23, 2005 and February 25, 2005.

Macomb, Alexander, *A Treatise on Martial Law and Courts Martial as Practices in the United States of America*, Charleston, SC, US Military Philosophical Society, 1809

Martin, Paul and Szuter, Christine, "War Zones and Game Sinks in Lewis and Clark's West," *Conservation Biology*, Feb 1999, vol 13 no 1, page 36-45.

Moore, Robert J. Jr. and Haynes, Michael. *Lewis and Clark; Tailor Made, Trail Worn*, Helena, Far Country Press. 2003

Moulton, Gary E. editor, *The journals of the Lewis and Clark Expedition*, 13 volumes, Lincoln, University of Nebraska Press, 1983

Mulholland, James, *A History of Metals in Colonial America*, Fayetteville, University of Arkansas Press, 1981

Phillips, Wayne H., *Plants of the Lewis and Clark Expedition*, Missoula, Mountain Press Publishing Co., 2003

Porter, C. L., *Taxonomy of Flowering Plants*, San Francisco, W. H. Freeman and Co., 1959

Ragsdale, John G. *Dutch Ovens Chronicled; their use in the United States*, Fayetteville, University of Arkansas Press, 1991

Ronda, James P., *Lewis and Clark Among the Indians*, Lincoln, University of Nebraska Press, 1984

Royer, France and Dickinson, Richard, *Weeds of the Northern U.S. and Canada*, Edmonton, University of Alberta Press, 1999

Sloan, Colleen, Personal interview, Great Falls, MT, January 31, 2004

Smithers, Jim. "The Realities and Complexities of Food for Sir Alexander MacKenzie (1789) and Canada Sea-to-Sea, *"We Proceeded On*, February 1989

Triber, Jayne E., *A True Republican; the life of Paul Revere*, Boston, University of Massachusetts Press, 1998

Walchek, Kenneth, "Encountering Feast and Famine on the Lewis and Clark Trail," unpublished manuscript, Lewis and Clark Trail Heritage Foundation archives, Great Falls, MT, February 6, 2003

Website: http://members.aol.com/Srlohnes/rations.html; "Soldiers Rations During the Early Years of the American Revolution," a research project of the 25th Continental Regiment, a living history group.

Website: www.osv.org article written by Frank G. White, January 1980

Website: www.rkymtnhi.com
Website: www.coffeenewsco.com
Website: www.garfoods.com
Website: http://oregontrail.blm.gov
Website: www.idos.com
Website: www.oldandsold.com
Website: www.richomdville.com
Website: www.dutchovenmagic.com
Website: www.alpinetrail.com
Website: http://home.earthlink.net/~rfrmac/Pages/dutch_info-2.html
Website: http://longhunter.tripod.com
Website: www.dutchovencookware.com
Website: www.wedlinydomowe.com/smoking-meat.html
Website: www.arkson.com/health/vbone.htm
Website: www.mayoclinic.com/health/foodpoisoning/ DS00981/DSECTION=symptoms
Whitney, and Rolfes, *Understanding Nutrition*, Belmont, CA, Wadsworth/Thompson Learning, 2002
Wilk, Kenneth, Army Corps of Engineers Lewis and Clark Bicentennial Assistant Co-ordinator, Personal correspondence November 17, 2003 and January 26, 2004.
Willard, Terry, *Edible and Medicinal Plants of the Rocky Mountains and Neighbouring Territories*, Calgary, Wild Rose College of Natural Healing, Ltd, 1992

Alphabetical Index

Ahsahka 128
Alderdale 126
Aldrich Point 121
American Flag 129, 142
Antelope 43, 44, 46, 47, 48, 51, 60, 80, 83, 84, 122, 151, 154, 233
Aricara 46, 47
Asimina Triloba 148, 233
Assiniboine 60
Augusta 136
Backwoodsmen 174
Bacteria 34, 95, 106, 108, 109, 110, 207
Baskets 46, 113, 114
Beads 20, 47
Beans 25, 32, 33, 34, 46, 47, 54, 57, 75, 87, 146, 163, 164, 169, 187, 217, 219, 234
Bear 13, 17, 23, 33, 34, 48, 60, 61, 63, 64, 65, 67, 68, 69, 72, 75, 77, 91, 96, 129, 130, 133, 137, 138, 139, 143, 144, 156, 174, 175, 201, 207, 208, 211, 219, 221, 226, 233
Beargrass 113, 114
Beaver 37, 39, 40, 41, 59, 60, 61, 64, 68, 82, 83, 86, 90, 111, 116, 136, 154, 156, 233
Beaverhead 5, 79, 82, 83, 136, 240
Beaverhead Rock 5, 79, 82, 83, 240
Beer 98
Big Hole River 136
Big Hole Valley 136
Big Horn 139
Bighorn Sheep 64, 154, 233
Billings 139
Biscuit 43, 74
Biscuits 25, 26, 157
Bisquits 35, 59, 234
Bissell 21
Bitterroot 87, 90, 135, 234, 236
Blackberries 110
Blackfoot River Valley 136
Blacksmithing 54

Blacksmiths 55, 56, 68, 164, 169, 170, 201, 206
Blubber 111, 112, 115, 156, 235, 236
Blue Camas 128
Blue grouse 89
Boiling 33, 34, 36, 37, 40, 41, 48, 57, 59, 60, 61, 62, 63, 76, 81, 106, 126, 130, 162, 165, 166, 169, 172, 183, 184, 185, 187, 188, 196, 201, 207, 208, 209, 237, 238
Boils 34, 38, 115
Bonneville Dam 124
Boudin blanc 63, 207, 225
Bozeman 137
Bradford Island 124
Brains 64, 76, 114
Brants 101
Brass 20, 24, 27, 72, 96, 97, 101, 125, 159, 160, 162, 165, 166, 180, 184, 194, 195, 196, 200, 204, 205, 215, 219
Brass kettles 20
Bratten 41
Bratton 40, 63, 110, 116, 118, 121, 122, 137
Bread 13, 24, 26, 32, 35, 47, 57, 62, 63, 73, 74, 76, 81, 92, 93, 95, 98, 101, 107, 117, 128, 148, 161, 172, 185, 186, 187, 212, 215, 227, 234, 236, 237
Bread 13
Breadroot 71, 145, 234
Brisket 116
British musket 100
Brownsmead 121
Brush drag 40
Buffalo 33, 40, 41, 42, 43, 44, 45, 48, 52, 53, 54, 55, 56, 57, 60, 61, 62, 63, 69, 70, 73, 74, 75, 76, 77, 80, 88, 92, 104, 134, 136, 137, 138, 139, 142, 143, 144, 145, 147, 151, 153, 154,

156, 157, 172, 174, 177, 185,
186, 187, 201, 202, 207, 208,
225, 233, 242, 248
Buffalo hump 63, 74
Buffalo meat 44, 45, 74, 142, 147,
157, 185, 186, 207
Burrowing squirrel 67
Butcher knives 20, 27
Camas 92, 93, 95, 96, 128, 131,
132, 155, 175, 176, 209, 217,
221, 227, 234, 236
Camas flats 131, 132
Camassia quamash 92, 128
Camawait 86
Cameahwait 87
Camp Chopunnish 129
Camp Disappointment 139, 219
Camp Dubois 5, 9, 16, 19, 22, 23,
24, 27, 31, 153, 154, 158, 159,
160, 161, 163, 169, 170, 177,
189, 191, 193, 201, 206, 207,
211, 218, 235, 239
Camp Fortunate 136
Cannon Beach 112
Canoe 5, 67, 73, 77, 79, 81, 96, 97,
101, 103, 106, 113, 118, 125,
126, 136, 142, 143, 147, 155,
160, 161, 193, 219, 240
Canoe Camp 5, 67, 77, 79, 155,
240
Canoes 56, 59, 71, 77, 79, 80, 81,
83, 84, 87, 95, 96, 99, 100,
101, 103, 104, 106, 107, 111,
112, 115, 116, 118, 119, 123,
124, 125, 126, 135, 136, 137,
138, 139, 142, 143, 144, 146,
149, 155, 160, 162, 197, 205
Cascades 100, 124
Cast iron 162, 164, 165, 166, 167,
168, 178, 179, 180, 187, 193,
194, 195, 196, 197, 218, 219
Cast iron kettle 166
Catfish 36, 37, 40, 42, 44, 64, 148,
156
Charbeneau 137

Charbonneau 61, 73, 74, 85, 87,
146, 201, 213, 225
Cheyenne 146
Chief Big White 153, 210
Chokecherries 54, 68, 233
Clark 1, 7, 9, 11, 12, 15, 16, 17, 20,
21, 22, 23, 24, 25, 26, 31, 33,
34, 35, 36, 37, 38, 39, 40, 41,
42, 43, 44, 45, 46, 47, 48, 52,
53, 54, 55, 56, 57, 61, 67, 68,
69, 70, 71, 72, 73, 74, 77, 79,
80, 81, 82, 83, 84, 85, 86, 87,
88, 89, 91, 92, 93, 95, 96, 97,
98, 99, 100, 101, 102, 103,
105, 106, 107, 108, 109, 110,
111, 112, 118, 121, 123, 124,
125, 126, 130, 135, 136, 137,
138, 139, 143, 144, 145, 146,
147, 148, 149, 154, 155, 156,
157, 159, 160, 163, 167, 169,
170, 174, 175, 176, 177, 184,
185, 186, 187, 191, 192, 193,
195, 199, 201, 203, 204, 206,
210, 211, 212, 213, 217, 223,
229, 230, 231, 232, 239, 242,
243, 244, 245, 246, 247, 248
Clatskanie 121
Clatsop 5, 12, 15, 106, 116, 118,
119, 121, 155, 156, 160, 161,
175, 178, 191, 201, 204, 205,
207, 208, 212, 215, 218, 219,
241
Clatsop Indian 116
Clearwater 95, 106, 127, 128, 129,
155, 175
Clearwater River 95, 127, 155, 175
Coffee 5, 24, 25, 36, 73, 159, 163,
164, 234
Coffee Grinders 5, 159, 163
Coffee mills 164
Collins 35, 41, 98, 110, 137, 231,
248
Colt 89, 90, 91, 127, 130, 137
Colter 39, 41, 42, 44, 88, 90, 137,
146

Columbia 81, 83, 86, 90, 97, 98, 99, 100, 105, 106, 117, 119, 121, 122, 126, 127, 135, 156, 193, 196, 208
Columbia river 81, 83, 90, 97, 99, 122
Conrad 142
Cooking utensils 5, 58, 159, 160, 163, 167, 168, 171, 197, 203, 231
Copper teakettles 100
Corn 5, 13, 24, 25, 26, 31, 32, 36, 39, 40, 41, 42, 46, 47, 48, 51, 52, 54, 55, 56, 57, 68, 86, 87, 88, 146, 147, 153, 154, 157, 159, 163, 164, 174, 186, 196, 230, 234, 237, 238
Corn meal 25, 26, 32, 36, 40, 55, 57, 68, 174, 196, 238
Corn mill 48
Corn mills 5, 26, 159, 163, 164
Cornmeal 153, 184, 196, 209, 215, 236, 238
Council Bluff 160
Court martials 35, 192
Cous Bisquit-Root 128
Cous roots 127, 129, 217, 219, 221
Cramps 34, 95
Crawfish 92
Cruzatte 117, 137, 145
Cruzzatte 48
Cut Bank Creek 138
Dahila, Washington 102
DeChamps 32
Deer 7, 13, 23, 31, 32, 33, 34, 35, 36, 37, 39, 41, 43, 44, 46, 48, 52, 53, 55, 56, 59, 60, 64, 67, 69, 76, 77, 79, 80, 81, 84, 85, 86, 88, 89, 90, 91, 95, 99, 100, 101, 102, 105, 107, 111, 119, 122, 123, 124, 127, 128, 130, 131, 132, 133, 135, 136, 146, 151, 153, 154, 155, 156, 172, 174, 177, 178, 233, 237, 242
Detachment Order 32, 169, 206, 228, 230, 231, 232

Detachment orders 153
Diet 5, 12, 13, 14, 15, 21, 24, 34, 37, 46, 61, 81, 95, 97, 98, 99, 100, 107, 110, 117, 124, 153, 154, 155, 156, 158, 160, 178, 185, 190, 202, 219, 222, 242
Dog 42, 45, 67, 96, 98, 100, 111, 121, 124, 126, 127, 155, 156, 157, 233, 234, 236
Dogs 43, 45, 52, 67, 85, 96, 97, 98, 99, 100, 102, 111, 116, 122, 124, 125, 126, 127, 155, 218, 219
Drouillard 108, 110, 117, 118, 132, 140, 141, 142, 161
Dried berries 83, 92
Dried meat 33, 41, 43, 45, 46, 70, 71, 73, 79, 80, 123, 124, 127, 138, 139, 142, 143, 144
Dried roots 62, 156
Drouillard 31, 32, 33, 34, 36, 39, 41, 61, 67, 72, 73, 81, 83, 84, 85, 86, 135, 137, 141
Duck 80, 233
Ducks 59, 97, 98, 100, 101, 102, 126, 215
Dugout canoes 56, 77, 99, 160, 162, 197, 205
Dugouts 70, 77, 81, 82, 83, 84, 85, 95, 103, 118, 197, 203
Dumpling 73, 74, 205
Dumplings 73, 74, 75, 76, 172, 187, 209, 216, 218, 236
Dutch oven 7, 41, 73, 74, 76, 159, 162, 165, 166, 169, 172, 178, 182, 187, 195, 196, 197, 198, 199, 205, 216, 217
Dutch Ovens 5, 7, 57, 159, 162, 164, 165, 166, 178, 179, 180, 181, 187, 193, 194, 196, 197, 198, 200, 205, 218, 242, 245
Dysentery 34, 38, 39, 73, 95, 207
Ecola Creek 112
Elk 37, 39, 42, 43, 44, 46, 48, 52, 55, 56, 60, 61, 64, 65, 67, 72, 73, 76, 77, 79, 80, 81, 90, 94,

249

102, 107, 108, 109, 110, 111, 112, 113, 114, 115, 116, 117, 118, 119, 121, 122, 123, 124, 126, 136, 143, 144, 145, 146, 148, 151, 153, 154, 155, 156, 162, 175, 177, 201, 205, 207, 208, 233, 235, 242
Elk fleece 37
Elk Tongues 110
Espontoon 69
Exploring the Big Sky 186
Fenel roots 130
Fever 34, 95
Field 5, 11, 19, 21, 31, 53, 55, 131, 135, 161, 183, 184, 185, 186, 188, 189, 191, 200, 211, 213, 221, 222, 233
Field tests 5, 183, 186, 189
Fields 41, 42, 44, 72, 73, 92, 93, 110, 137, 140, 141, 142
Fire irons 5, 159, 169, 170, 171, 187, 193, 201
Fire wood 97, 127
Fireplace crane 218
Firewood 121, 125, 126, 185, 206, 213, 231
Fish 20, 21, 36, 37, 40, 42, 47, 61, 64, 69, 70, 71, 86, 87, 88, 89, 93, 95, 96, 97, 98, 99, 101, 102, 103, 106, 107, 109, 110, 113, 114, 117, 118, 121, 122, 125, 132, 153, 156, 161, 164, 172, 186, 187, 212, 215, 216, 217, 219, 233, 234, 236
Fish hooks 20, 47, 110
Fishing net 96
Flathead Indians 89
Fleas 102, 110
Flintlock rifles 60
Flour 24, 25, 26, 32, 39, 41, 45, 54, 57, 62, 63, 68, 74, 76, 79, 81, 83, 84, 88, 99, 142, 147, 148, 153, 154, 157, 172, 184, 196, 209, 215, 225, 230, 234, 236
Floyd 15, 32, 39, 40, 50, 104, 148, 228, 230, 231

Food preparation 14, 16, 184, 190, 192, 221
Fort 5, 12, 15, 35, 49, 51, 52, 53, 54, 55, 56, 57, 58, 108, 110, 111, 112, 115, 116, 117, 119, 121, 130, 136, 137, 142, 143, 146, 153, 155, 156, 160, 161, 175, 178, 191, 192, 193, 199, 201, 204, 205, 207, 208, 212, 215, 222, 236, 241, 244, 247, 248
Fort Clatsop 12
Fraser 41
Frazer 135
Fresh meat 19, 23, 24, 25, 31, 32, 41, 46, 51, 52, 54, 55, 56, 59, 77, 79, 80, 81, 84, 87, 88, 89, 110, 112, 117, 122, 143, 153, 154, 207, 230, 231
Fry bread 74, 81, 148, 185, 215
Frying 5, 36, 57, 73, 74, 159, 162, 164, 166, 167, 168, 169, 170, 180, 195, 196, 197, 205, 207, 209, 217, 219, 236
Frying pan. 205, 217
Frying pans 5, 159, 162, 167, 168, 169, 180, 196, 197, 205
Frying pans 196
Gairdner's yampah 88
Gallitan 137
Gass 15, 40, 42, 51, 76, 90, 92, 95, 96, 99, 101, 105, 111, 116, 129, 135, 142, 143, 148
Geese 23, 37, 59, 60, 79, 97, 101, 156, 233
Gibson 40, 110, 116, 137, 138, 244
Glendive 143, 144
Goodrich 36, 69, 70, 135
Goose 41, 59, 61, 99, 100, 215
Grease 25, 32, 33, 34, 37, 41, 57, 59, 61, 62, 63, 64, 73, 75, 85, 89, 90, 96, 99, 105, 107, 112, 125, 132, 133, 138, 143, 144, 164, 175, 184, 185, 201, 207, 215, 217, 230, 236

Great Falls 65, 67, 69, 70, 76, 77, 78, 122, 135, 136, 137, 156, 157, 184, 205, 207, 208, 213, 219, 243, 245, 247, 248
Grizzly bear 48, 61, 63
Gros Ventres 48
Gun 48, 69, 86, 91, 113, 125, 137, 140, 141, 142
Guns 21, 113, 139, 140, 141, 143, 154
Hail 108
Hail stones 74
Hall 35, 137
Hamilton 135
Hidatsa 48, 89, 90, 145
Hog lard 25, 32, 186
Homer 148
Honor Guard 7, 184, 185, 186, 189, 247, 248
Horse 40, 44, 53, 85, 86, 90, 91, 92, 96, 97, 124, 125, 126, 127, 128, 129, 130, 131, 138, 141, 155, 234
Horses 34, 52, 53, 56, 84, 86, 87, 88, 89, 90, 92, 95, 96, 97, 99, 122, 123, 124, 125, 126, 127, 128, 129, 130, 131, 132, 133, 135, 136, 137, 139, 140, 141, 142, 145, 154
Horseshoe Point 143
Hot Springs at Lolo 133
Howard 42, 137
Hunter 13, 20, 23, 25, 36, 113, 174, 209, 210, 212
Hunters 20, 84, 118, 153, 156, 178
Hunting 7, 11, 13, 19, 21, 22, 23, 25, 32, 40, 42, 44, 46, 51, 53, 54, 55, 57, 58, 60, 67, 71, 76, 80, 81, 84, 90, 91, 99, 107, 108, 112, 113, 116, 118, 123, 130, 131, 136, 153, 154, 157, 173, 174, 190, 208, 209, 212, 220, 242, 244, 248
Idaho 127, 128, 129, 131, 155, 175, 176, 178, 216, 217
Indian woman 15, 59

Iron boat 65, 67, 72, 73, 75, 76, 205
Iron tripods 162, 193, 194, 206
Irvine 26
Jefferson 14, 19, 20, 22, 25, 43, 57, 59, 80, 81, 135, 136, 137, 159, 161, 162, 174, 176, 190, 193, 206, 223, 243, 244
Jerked 33, 34, 37, 41, 42, 43, 44, 70, 71, 114, 116, 154
Jessomme 146
John Day River 99, 106
Joseph 41, 42, 44, 72, 73, 92, 93, 110, 137, 140, 141, 142
Kamiah 129
Kansas River 34
Karlson Island 121
Kaskaskia 194, 199
Keelboat 23, 24, 25, 57, 161, 193, 204
Kettle 41, 54, 72, 73, 96, 125, 160, 161, 166, 167, 169, 172, 183, 184, 186, 187, 188, 194, 195, 196, 204, 205, 206, 208, 212, 215, 217, 218, 219, 225
Kettles 16, 20, 24, 27, 37, 41, 42, 81, 89, 95, 96, 97, 98, 99, 110, 115, 116, 125, 159, 160, 162, 164, 165, 166, 167, 170, 171, 180, 184, 187, 194, 195, 196, 200, 201, 203, 204, 205, 218, 220, 228
Labich 110, 114, 117, 139
LaBiche 41, 137
Labiech 101
LaCharette 149
Lead canisters 115
Leather teepee 130
LePage 137
Lewis 1, 7, 9, 11, 12, 15, 17, 20, 21, 22, 23, 26, 27, 32, 33, 35, 36, 39, 40, 41, 42, 43, 45, 47, 53, 55, 56, 59, 60, 61, 62, 63, 64, 65, 67, 68, 69, 70, 71, 72, 73, 74, 75, 76, 77, 78, 79, 80, 81, 83, 84, 85, 86, 87, 88, 89, 90, 92, 93, 95, 96, 100, 105, 107,

108, 110, 111, 112, 113, 114,
115, 116, 117, 118, 121, 122,
123, 124, 125, 127, 128, 129,
130, 131, 132, 133, 135, 136,
137, 138, 139, 140, 141, 142,
143, 144, 145, 147, 149, 155,
156, 157, 160, 161, 162, 163,
164, 165, 167, 168, 169, 170,
174, 175, 177, 184, 185, 186,
187, 189, 191, 192, 193, 194,
196, 197, 199, 201, 203, 204,
205, 206, 207, 212, 213, 217,
219, 221, 223, 225, 226, 227,
229, 230, 231, 232, 239, 242,
243, 244, 245, 246, 247, 248
Lewis 20
Livingston 137
Lomatium Cous 128, 234
Mandan 5, 19, 22, 24, 47, 48, 51,
57, 59, 68, 75, 130, 139, 145,
146, 147, 153, 159, 161, 177,
191, 201, 204, 205, 210, 211,
212, 215, 218, 219, 222, 236,
239
Mandans 22, 87, 139, 147, 157,
210
Marias 5, 59, 67, 68, 135, 137, 138,
142, 143, 147, 154, 156, 157,
162, 165, 186, 187, 196, 197,
205, 219, 239, 240, 247
Marias 219
Marias River 5, 67, 68, 135, 137,
138, 142, 143, 147, 154, 157,
162, 165, 186, 205, 240
Marrow 45, 57, 60, 62, 67, 107,
110, 112, 116, 237
Marrow bones 60, 67, 107, 110,
112, 116
Massac 194, 199
McNeal 83, 84, 135
McNeil 83, 84, 85, 137
Medicine River 70, 139, 140
Menetares 48
Mess 22, 24, 25, 32, 35, 36, 37, 39,
42, 71, 125, 133, 167, 185,
196, 201, 205, 210, 211, 212,
213, 215, 228, 230, 231, 232
Mess gear 167
Military unit 5, 11, 19, 20, 21, 23,
24, 159, 161, 162, 163, 190,
191, 192, 193, 195, 196, 197,
200, 201, 211, 223
Mill Creek 99, 124
Minnetaries 139
Missouri 5, 19, 24, 25, 27, 31, 32,
38, 39, 44, 46, 48, 56, 57, 59,
62, 65, 67, 68, 69, 70, 72, 76,
77, 78, 80, 83, 88, 105, 116,
123, 135, 136, 137, 138, 142,
144, 145, 146, 148, 153, 157,
161, 185, 193, 194, 195, 196,
197, 207, 219, 225, 235, 239
Missouri River 19, 25, 31, 88, 157,
185, 195
Moccasins 67
Montana 15, 27, 34, 36, 44, 46, 53,
59, 84, 96, 97, 100, 103, 114,
125, 132, 135, 136, 137, 139,
140, 141, 142, 144, 153, 155,
156, 157, 158, 167, 173, 174,
177, 178, 185, 193, 202, 208,
210, 213, 221, 243, 244, 247,
248
Mortar and pestle 164
Mosquito 33
Mosquito repellant 33
Mosquitoes 34, 59, 74, 144
Multnomah Falls 124
Muscle aches 34
Mush 138, 209, 217, 219, 236
Musselshell River 64, 143
Newman 42, 47
Nez Perce 5, 83, 92, 95, 127, 131,
155, 156, 157, 226, 227
Nez Perces 5, 90, 97, 129, 175,
176, 240
North Dakota 156, 202
Old Toby 89
Onion 79, 130, 227
Onions 60, 79, 130, 233, 234

Ordway 15, 32, 36, 41, 42, 45, 46, 57, 74, 77, 86, 89, 90, 92, 109, 116, 118, 135, 137, 138, 139, 142, 143, 149, 160, 213, 228, 229, 230, 231
Oregon 157
Otter 127, 233
Pacific Ocean 5, 19, 84, 95, 104, 105, 155, 240
Packer Meadows 133
Palouses. 97
Papaws 157
Park City 138
Passmore 26
Paul Revere 5, 165, 178, 179, 180, 181, 197, 198, 199, 243, 244, 245
Paul Revere 243
Paw Paws 148, 149
Pawpaws 148
Peas 25, 32, 33, 34, 92, 217, 234
Pelican 157, 233
Pemmican 45, 58, 75, 77, 79, 83
Pheasants 92, 187
Pheasants, 92
Pheasent 99
Philadelphia 20, 26, 27, 160, 161, 162, 163, 164, 168, 194, 200, 203, 204, 206
Pillar Rock 102
Pipe tomahawk 101
Pirogue 47, 51, 52, 64, 70, 142, 143, 144, 193, 203
Pirogues 25, 31, 32, 42, 56, 59, 193, 197, 232
Pit cooking. 175
Plums 36, 43, 157, 233
Point Adams 110
Pomp 137, 144, 146
Pork 25, 26, 32, 39, 44, 46, 57, 60, 61, 68, 83, 86, 89, 142, 153, 154, 157, 230, 234
Portable soup 6, 26, 57, 90, 91, 92, 154, 234, 236, 237
Portage 71, 72, 74, 77, 99, 100, 101, 124, 137, 139, 185, 186, 187, 193, 196, 197, 201, 205, 207, 208, 216, 217, 247, 248
Portage Route Chapter 247, 248
Portland 101, 122
Potlatch River 96
Potts 137
Prairie dog 67, 157, 233, 236
Prairie Larks 41
Prairie onions 60
Prarie woolf 92
Pronghorn 43
Provisions 22, 24, 25, 26, 32, 35, 39, 59, 68, 87, 92, 98, 106, 117, 119, 138, 153, 154, 156, 186, 196, 204, 206, 209, 230, 231
Pryor 32, 52, 55, 137, 145, 228, 230, 231
Psoralea esculenta 45, 62, 71
Pumpkins 46, 51
Purchase orders 20
Pyror 139
Rabbits 23, 60, 157, 233
Ragsdale 165, 178, 180, 197, 198, 242, 243, 245
Rations 11, 21, 22, 24, 26, 29, 31, 32, 34, 35, 42, 51, 53, 54, 55, 59, 68, 85, 151, 159, 172, 209, 217, 245
Rattlesnake Creek 135
Ravens 56
Revere 243
Ricaras 146
Roasting 33, 36, 39, 41, 45, 46, 57, 59, 61, 67, 75, 89, 90, 100, 102, 105, 114, 130, 159, 162, 163, 164, 170, 196, 202, 207, 209, 216, 236
Rocky Mountains 64, 93, 122, 123, 154, 162, 175, 196, 197, 201, 222
Rooster Rock 101
Roots 13
Ruben 35, 41, 42, 44, 72, 73, 92, 93, 98, 110, 137, 140, 141, 142, 231, 248

Rushes Pills 95, 96
S 157
Sacagawea 59, 62, 70, 71, 74, 80, 85, 86, 87, 97, 106, 107, 111, 131, 135, 137, 146
Sacajawea 154, 173, 175, 176, 183, 210, 234, 235
Salmon 70, 83, 84, 86, 87, 88, 89, 93, 95, 96, 97, 98, 99, 100, 103, 104, 105, 106, 107, 109, 122, 124, 125, 126, 127, 129, 130, 131, 155
Salmon River 86, 87, 88, 89
Salmonella 95
Salt 24, 25, 26, 32, 45, 57, 63, 68, 74, 75, 76, 81, 90, 95, 102, 105, 106, 107, 108, 109, 110, 111, 112, 113, 114, 115, 116, 119, 132, 142, 155, 157, 160, 174, 183, 225, 234, 237, 238
Salt makers 111, 112, 115, 160
Sandy River 122
Saucepan 27
Sauger 69
Saugers 69
Scaffolds 97, 123, 227
Scriver 1, 2, 247, 248
Seal 52, 76, 103, 121
Serviceberry Valley 85
Shannon 42, 43, 72, 73, 108, 110, 111, 114, 117, 132, 137, 139, 228
Sheep 64, 79, 81, 143, 154, 233
Sheilds 137
Shield 32
Shields 41, 42, 54, 177
Ships 101, 105, 119
Shoshone 79, 80, 82, 84, 88, 89, 90, 175
Sioux 42, 44, 45, 46, 47, 146, 147, 160
Snake River 96, 127
Sockeye 103
South Dakaota 157
South Dakota 147, 175

Spit 7, 24, 75, 114, 170, 171, 187, 201, 206, 216, 217, 218, 219
Spits 113, 162, 196, 206, 217
Spoons 97, 113, 114, 163, 167, 168, 171, 194, 200, 201, 203, 204, 205, 217
Squash 46, 47, 48, 51, 57, 58, 87, 146, 153, 185, 187, 217, 235, 236
Squirrels 67, 99, 131
St Louis 59, 116, 148, 149
St. Louis 16, 22, 23, 24, 25, 27, 51, 59, 119, 153, 154, 157, 159, 160, 161, 190, 194, 195, 199, 200, 201, 203, 204, 206, 209, 217, 222, 232
Steelhead trout 99
Sturgeon 101, 117, 118, 121, 122
Suet dumpling 74, 205
Sugar 24, 25, 37, 87, 157, 234
Sulfur spring 71
Sun flowers 88
Sun Rivers 136
Superintendents of Provisions 35, 186, 206, 209, 231
Superintendents of Provisions 186
Sword 45, 100, 126
Tarts 41, 209, 216, 217, 236
Teton River 44, 137, 142
The Dalles 99, 124
The River of no Return 89
Theories 5, 8, 16, 17, 183, 184, 199, 200, 221
Thirteen Mile Creek 144
Thompson 35, 39, 135, 231
Three Forks of the Missouri 80
Tobacco 42, 46, 47, 87, 109, 118, 136, 147, 235, 236
Tobacco root 87, 235, 236
Toenyes 1, 2, 185, 247, 248
Tomahawk 54, 101, 124
Trading 20, 25, 51, 55, 56, 89, 95, 106, 112, 113, 116, 117, 124, 125, 131, 155, 159, 194, 206, 209, 235

Travelers Rest 5, 90, 121, 132, 133, 135, 156, 213, 241, 242
Travelers Rest 241
Travelers Rest 90, 132, 133
Trout 40, 70, 84, 86, 87, 99, 157, 185, 188, 219, 221
Turkey 23, 39, 43, 59, 157, 186, 210, 234
Twin Bridges 136
Two Medicine River 139, 140
Umatilla 97
Umatillas 126
Vancover 122
Venison 37, 59, 69, 80, 81, 83, 85, 105, 124, 128, 130, 133
Walla Walla 126, 127
Walula 97
Wanapams 97
War Department 26, 161, 191, 192, 194
Warfinton 204
Warvington 32, 228
Washington 102, 122, 126, 157, 195, 200, 217, 242
Watermelons 39
Weippe 131, 155, 176
Weiser 117, 137, 231, 248
Werner 35, 135
Western Spring Beauty 87
Whale 15, 110, 111, 112, 115, 156, 235, 236
Whiskey 26, 32, 35, 40, 46, 148, 149
White apple 45, 62, 145, 234
White Bear Island 72, 137, 138, 139
White Cliffs 65, 143
Whitehouse 15, 52, 55, 73, 75, 81, 83, 87, 91, 95, 96, 101, 104, 106, 109, 115
Wild berries 13
Willard 31, 36, 54, 110, 111, 116, 121, 122, 125, 137, 142, 143
Williamette River 123, 127
Willow run 73, 74
Windsor 137, 139
Wiser 39, 110, 111
Wolf 36, 55
Wolves 53, 56, 68, 69, 72, 91, 94, 139, 177
Wooden bowls 113
Wooden spoons 113
Yakima's 97
Yampah 88, 175, 176, 234, 236
Yankton 147
Yellowstone 60, 135, 137, 138, 139, 143, 144, 146, 156, 213
York 33, 40, 41, 43, 47, 54, 74, 106, 137, 201, 210, 213, 239, 243, 244
Yucca root 183
Myths 5, 166, 173, 242

The Authors

Phil Scriver

John Toenyes

This book is the result of many years of experience by the authors, John Toenyes and Phil Scriver. Both are long time Lewis and Clark buffs who have actively participated in many projects; have spent years re-creating the Lewis and Clark story; have read a wide variety of related books and articles; and have given many presentations and skills demonstrations to school classes, civic organizations and tourist groups.

Phil Scriver has been immersed in history and writing most of his adult life. He has chaired two projects to create outdoor bronze statues of Lewis and Clark, specializing in the historical research and promotions for both. "The Explorers at the Marias" statue was dedicated in Fort Benton, Montana on June 13, 1976 and "The Explorers at the Portage" was dedicated in Great Falls, Montana on July 4, 1989.

He has been a member of the Lewis and Clark Trail Heritage Foundation since 1975 and of the Great Falls, Montana based Portage Route Chapter of the Foundation since 1988, serving on the Board of Directors for 14 years as President, Secretary and Director. Scriver has been a member of the Lewis and Clark Honor Guard since 1996. The Honor Guard is an organization for those who re-create the Lewis and Clark Expedition. They make and wear historically accurate clothes of the era, use historically accurate equipment and skills to present a period of time in the life of the Expedition's journey across the continent.

He served as a technical consultant for two documentaries on the Lewis and Clark Expedition. In 2003 Phil published his book "Lewis and Clark Passed Here" and in 2008 published his second book "Wrightings."

Scriver began his writing career as a reporter for the weekly newspaper in Fort Benton, The River Press. He wrote various news articles along with local and regional historical pieces. Eventually he focused his writing on history which led to publication of several articles in the Lewis and Clark Trail Heritage Foundation's quarterly publication, We Proceeded On.

In 1999 Phil developed the application for a Lewis and Clark bicentennial "signature event." His proposed event, one of 15 events to focus nationwide attention on specific areas of the route of Lewis and Clark's journey, would be held in the Fort Benton/Great Falls area. His application established the basis for all historical events and activities to be done during the 34 days of the signature event.

John Toenyes graduated from Montana State University with a Bachelor of Science degree in Entomology. He was in business in the environmental management industries then later purchased a full service restaurant. He spent 23 years in the food business consequently he understands the health and care of food products as well as their preparation and quality. He has been an outdoor cooking enthusiast for over 35 years, cooking for groups using authentic historic equipment and procedures. He has been a cook on a wagon train where he used a chuck wagon as well as for 25 years cooked for hunting camps, bus tours and of course Lewis and Clark historic re-enactments.

He was the planning cook for the Lewis and Clark bicentennial signature event historic re-enactment of the portage of the Great Falls. The historically authentic camp was fed the exact type of food that Lewis and Clark had available and it was cooked the same way they would have done it.

Toenyes worked as a location finder and camp co-ordinator in the Great Falls area for Ken Burns and Dayton Duncan during their filming of the Public Broadcasting System special documentary "Lewis and Clark Corps of Discovery." During the documentary filming Toenyes assisted in building authentic camps and cook fires utilizing original foods such as buffalo.

During the last 23 years Toenyes has been president of the Lewis and Clark Honor Guard twice and has served two terms on the Portage Route Chapter Board of Directors; he currently serves as vice president of the Chapter. He is very active in the living history of the Corps of Discovery and enthralled by details of the survival of the Expedition.

As members of the Lewis and Clark Honor Guard the authors portray two of the Expedition's cooks: John Toenyes as Private Peter Weiser and Phil Scriver as Private John Collins. Their diverse backgrounds have uniquely prepared them to collaborate on this book dealing with how the Corps of Discovery fed itself and survived during their 28-month journey.

Made in the USA
Charleston, SC
18 June 2010